QA 76.9 .A25 T43 1997

NEW ENGLAND INSTITUTE
OF TECHNOLOGY
LEARNING RESOURCES CENTER

Technology and Privacy: The New Landscape

Technology and Privacy: The New Landscape

edited by Philip E. Agre and Marc Rotenberg

NEW ENGLAND INSTITUTE
OF TECHNOLOGY
LEARNING RESOURCES CENTER

The MIT Press
Cambridge, Massachusetts
London, England

4\98   # 36485899

© 1997 Massachusetts Institute of Technology

All rights reserved. No part of this book may be reproduced in any form by any electronic or mechanical means (including photocopying, recording, or information storage and retrieval) without permission in writing from the publisher.

Set in Sabon by The MIT Press.
Printed and bound in the United States of America.

Library of Congress Cataloging-in-Publication Data

Technology and privacy : the new landscape / edited by Philip E. Agre and Marc Rotenberg.
   p.   cm.
Includes bibliographical references and index.
ISBN 0-262-01162-x (alk. paper)
1. Computer security. 2. Data protection. 3. Privacy, Right of. I. Agre, Philip. II. Rotenberg, Marc.
QA76.9.A25T43   1997
323.44′83—dc21   97-7989
                                                          CIP

Contents

Preface   vii

Introduction   1

1 Beyond the Mirror World: Privacy and the Representational
   Practices of Computing   29
   Philip E. Agre

2 Design for Privacy in Multimedia Computing and Communications
   Environments   63
   Victoria Bellotti

3 Convergence Revisited: Toward a Global Policy for the Protection
   of Personal Data?   99
   Colin J. Bennett

4 Privacy-Enhancing Technologies: Typology, Critique, Vision   125
   Herbert Burkert

5 Re-Engineering the Right to Privacy: How Privacy Has Been
   Transformed from a Right to a Commodity   143
   Simon G. Davies

6 Controlling Surveillance: Can Privacy Protection Be Made
   Effective?   167
   David H. Flaherty

7 Does Privacy Law Work?   193
   Robert Gellman

8 Generational Development of Data Protection in Europe   219
   Viktor Mayer-Schönberger

9   Cryptography, Secrets, and the Structuring of Trust   243
David J. Phillips

10 Interactivity As Though Privacy Mattered   277
Rohan Samarajiva

List of Contributors   311
Index   313

Preface

Ever since computers were first applied in organizational settings, nearly 50 years ago, social theory and popular imagination have associated information technology with the practices of bureaucratic control. Technological threats to personal privacy, accordingly, have been understood as the inexorable progress of Big Brother's agenda of total surveillance over our lives. Although the danger of surveillance technologies continues to grow and even accelerate, other innovations have invalidated the automatic identification between technology and control. These innovations include technologies of privacy protection that permit information systems to operate without identifying the individuals whose lives they affect; they also include communications technologies such as the Internet that make possible the construction of a new public sphere for debate and action around privacy issues.

As a result of these developments, the whole landscape of issues around technology and privacy has been transformed. The authors in this collection survey this new landscape. Each author brings a distinctive disciplinary background and personal experience, and each chapter contributes a conceptual framework for understanding the complex interactions among technology, economics, and policy through which the new landscape has evolved. Although the details of technology will surely change over the coming years, we hope that these analyses will provide continuing guidance by identifying some of the underlying trends.

We wish to thank the authors for their cooperation throughout the preparation of this volume, Paul Bethge for his highly competent editing, and Bob Prior for his editorial guidance and expert setting of deadlines.

Technology and Privacy: The New Landscape

Introduction

Philip E. Agre

Our premise in organizing this volume is that the policy debate around technology and privacy has been transformed since the 1980s. Tectonic shifts in the technical, economic, and policy domains have brought us to a new landscape that is more variegated, more dangerous, and more hopeful than before. These shifts include the emergence of digital communication networks on a global scale; emerging technologies for protecting communications and personal identity; new digital media that support a wide range of social relationships; a generation of technologically sophisticated privacy activists; a growing body of practical experience in developing and applying data-protection laws; and the rapid globalization of manufacturing, culture, and the policy process. Our goal is to describe this emerging landscape. By bringing together perspectives from political science, law, sociology, communications, and human-computer interaction, we hope to offer conceptual frameworks whose usefulness may outlive the frenetically changing details of particular cases. We believe that in the years ahead the public will increasingly confront important choices about law, technology, and institutional practice. This volume offers a starting point for analysis of these choices.

The purpose of this introduction is to summarize and synthesize the picture of this new landscape that the contributors have drawn. First, however, I should make clear what we have not done. We have not attempted to replace the foundational analysis of privacy that has already been admirably undertaken by Allen (1988), Schoeman (1984), and Westin (1967). We have not replicated the fine investigative work of Burnham (1983) and Smith (1979). Nor, unlike Lyon and Zureik (1996), have we tried to place the issues of privacy and surveillance in their broadest sociological context. Our work is organized conceptually and not by area of concern (medical, financial, marketing, workplace, political repression, and so on). Although our case studies are drawn from several countries, our method is not systematically comparative (see Bennett 1992, Flaherty 1989, and Nugter 1990). We have not attempted a complete survey of the issues that fall in the broad intersection of "technology" and "privacy." By "technology," for example, we mean information and communication technology; we do not address the concerns raised by biological technologies such as genetic analysis (Gostin 1995). Our concern with the interactions among technology, economics, and policy complements Smith's (1994) study of organizational issues and Regan's (1995) more detailed analysis of the legislative process. Nor, finally, do we provide a general theory of privacy or detailed policy proposals. We hope that our work will be helpful in framing the new policy debate, and we have analyzed several aspects of the development of privacy policy to date.

The New Landscape

Mayer-Schönberger's chapter describes the configuration of technology and privacy issues in the late 1960s and the early 1970s. In that period, privacy concerns focused on a small number of large centralized databases; although instrumental to the construction of the modern welfare state, these databases also recalled the role of centralized files in the fascist era. In the United States, concerns about privacy were raised in popular works by Ernst and Schwartz (1962), Brenton (1964), and Packard (1964) and in a detailed scholarly treatment by Westin (1967). In each case, though, the general form of the response was the same: an enforceable code of practice, which came to be known as "data protection" in Europe and as "privacy protection" in the United States. The premise underlying the Code of Fair Information Practices was the same in both places: organizations that collected personal information about individuals had certain responsibilities, and individuals had rights against organizations in possession of personal information. In some instances, these practices were codified by professions or industry associations. In other instances they were reduced to law. As a general matter, the focus was the centralized collection of data, specified in place and time, and under the specific responsibility of a known individual or organization. (These principles and their implementation are described by Gellman. I will use the term "data protection" here.) Data protection does not seek to influence the basic architecture of computer systems. Instead, it abstracts from that architecture to specify a series of policies about the creation, handling, and disposition of personal data. Mayer-Schönberger, Bennett, and Flaherty describe the subsequent evolution of the data-protection model. This model is by no means obsolete, but the world to which it originally responded has changed enormously.

Some of these changes are technical. Databases of personal information have grown exponentially in number and in variety. The techniques for constructing these databases have not changed in any fundamental way, but the techniques for using them have multiplied. Data-mining algorithms, for example, can extract commercially meaningful patterns from extremely large amounts of information. Market-segmentation methods permit organizations to target their attention to precisely defined subgroups (Gandy 1993). Contests, mass mailings, and other promotions are routinely organized for the sole purpose of gathering lists of individuals with defined interests. More data is gathered surreptitiously from individuals or sold by third parties.

The pervasive spread of computer networking has had numerous effects. It is now easier to merge databases. As Bennett observes, personal information now routinely flows across jurisdictional boundaries. Computer networking also provides an infrastructure for a wide variety of technologies that track the movements of people and things (Agre 1994). Many of these technologies depend on digital wireless communications and advanced sensors. Intelligent transportation systems, for example, presuppose the capacity to monitor traffic patterns across a broad geographic area (Branscomb and Keller 1996). These systems also exemplify the spread of databases whose contents maintain a real-time correspondence to the real-world circumstances that they represent. These computerized mediations of personal identity have become so extensive that some authors speak of the emergence of a "digital persona" that is integral to the construction of the social individual (Clarke 1994).

Computer networking also provides the basis for a new generation of advanced communication media. In the context of the analog telephone system, privacy concerns (e.g., wiretapping and the abuse of records of subscribers' calls) were largely circumscribed by the system's architecture. Newer media, such as the Internet and the online services discussed by Samarajiva, capture more detailed information about their users in digital form. Moreover, the media spaces that Bellotti describes are woven into their users' lives in a more intimate way than older communication technologies. Digital technology also increases both the capacity of law-enforcement authorities to monitor communications and the capacity of subscribers to protect them.

At the same time, the new media have provided the technical foundation for a new public sphere. Privacy activists and concerned technologists have used the Internet to organize themselves, broadcast information, and circulate software instantaneously without regard to jurisdictional boundaries. Low-cost electronic-mail alerts have been used in campaigns against consumer databases, expanded wiretapping capabilities, and government initiatives to regulate access to strong cryptography. Public-policy issues that would previously have been confined to a small community of specialists are now contested by tens of thousands of individuals. Although the success of these tactics in affecting policy decisions has not yet been evaluated, the trend toward greater public involvement has given the technology a powerful symbolic value.

Potentially the most significant technical innovation, though, is a class of privacy-enhancing technologies (PETs). Beginning with the publication of the first public-key cryptographic methods in the 1970s, mathematicians have constructed a formidable array of protocols for communicating and conducting transactions while controlling access to sensitive information. These techniques have become practical enough to be used in mass-market products, and Phillips analyzes some of the sharp conflicts that have been provoked by attempts to propagate them. PETs also mark a significant philosophical shift. By applying advanced mathematics to the protection of privacy, they disrupt the conventional pessimistic association between technology and social control. No longer are privacy advocates in the position of resisting technology as such, and no longer (as Burkert observes) can objectives of social control (if there are any) be hidden beneath the mask of technical necessity. As a result, policy debates have been opened where many had assumed that none would exist, and the simple tradeoff between privacy and functionality has given way to a more complex tradeoff among potentially numerous combinations of architecture and policy choices.

Other significant changes are political and economic. The data-protection model has matured. Privacy commissioners such as Flaherty have rendered hundreds of decisions in particular cases, and the nature and the limitations of the privacy commissioner's role have been clarified. It has become possible to ask how the effectiveness of privacy policies might be evaluated, although (as both Flaherty and Bennett observe) few useful methods have emerged for doing so. Pressures have arisen to tailor data-protection laws to the myriad circumstances in which they are applied, with the result that sectoral regulation has spread. In the United States, as Gellman observes, the sectoral approach has been the norm by default, with little uniformity in regulatory conception or method across the various industries. In most other industrial countries, by contrast, sectoral regulation has arisen through the adaptation and tailoring of a uniform regulatory philosophy.

This contrast reflects a deeper divide. Bennett describes the powerful forces working toward a global convergence of the conceptual content and the legal instruments of privacy policy. These forces include commonalities of technology, a well-networked global policy community, and the strictures on cross-border flows of personal data in the European Union's Data Protection Directive. Though the United States has moved slowly to establish formal privacy mechanisms and standardize privacy practices over the last two decades, it now appears that the globalization of markets, the growing pervasiveness of the Internet, and the implementation of the Data Protection Directive will bring new pressures to bear on the American privacy regime.

The evolution of privacy policy, meanwhile, has interacted with individual nations' political philosophies. Mayer-Schönberger argues that this interaction should be viewed not on a nation-by-nation basis but rather as the expression of a series of partial accommodations between the uniform regulation of data handling and liberal political values that tend to define privacy issues in terms of localized interactions among individuals.

(This tension runs throughout the contemporary debate and will recur in various guises throughout this introduction.)

One constant across this history is the notorious difficulty of defining the concept of privacy. The lack of satisfactory definitions has obstructed public debate by making it hard to support detailed policy prescriptions with logical arguments from accepted moral premises. Attempts to ground privacy rights in first principles have foundered, suggesting their inherent complexity as social goods. Bennett points out that privacy is more difficult to measure than other objects of public concern, such as environmental pollution. The extreme lack of transparency in societal transfers of personal data, moreover, gives the issue a nebulous character. Citizens may be aware that they suffer harm from the circulation of computerized information about them, but they usually cannot reconstruct the connections between cause and effect. This may account in part for the striking mismatch between public expression of concern in opinion polls and the almost complete absence of popular mobilization in support of privacy rights.

One result of this unsatisfactory situation is that the debate has often returned to the basics. Mayer-Schönberger and Davies both remark on the gap between the technical concept of data protection and the legal and moral concept of privacy, but they assign different significance to it. For Mayer-Schönberger, the concept of data protection is well fitted to the values of the welfare state. Davies, however, focuses on the range of issues that data protection appears to leave out, and he regards the narrowly technical discourse of data protection as ill suited to the robust popular debate that the issues deserve.

The basic picture, then, is as follows: Privacy issues have begun to arise in more various and more intimate ways, a greater range of design and policy options are available, and some decisions must therefore be made that are both fundamental and extraordinarily complicated. Perhaps the most basic of these strategic decisions concerns the direction of technical means for protecting privacy. One approach, exemplified by Bellotti's study, is to provide individuals with a range of means by which to control access to information about themselves. Another approach, discussed in detail by Phillips and by Burkert, is to prevent the abuse of personal information from the start through the application of privacy-enhancing technologies that prevent sensitive data from being personally identifiable. PETs may also be appealing because, unlike privacy codes, they are to some extent self-enforcing. Although PETs may not eliminate the need for ongoing regulatory intervention, they seem likely to reduce it.

Negotiated Relationships

Ideas about privacy have often been challenged by new technologies. The existing ideas arose as culturally specific ways of articulating the interests that have been wronged in particular situations. As a result, cultural ideas about privacy will always tacitly presuppose a certain social and technological environment—an environment in which those kinds of wrongs can occur and in which other kinds of wrongs are either impossible, impractical, or incapable of yielding any benefit to a wrongdoer. As new technologies are adopted and incorporated into the routines of daily life, new wrongs can occur, and these wrongs are often found to invalidate the tacit presuppositions on which ideas about privacy had formerly been based. The development of law illustrates this. In a landmark 1967 case, *Katz v United States* (389 US 347), the US Supreme Court found that a warrantless police recording device attached to the outside of a telephone booth violated the Fourth Amendment's protection against unreasonable searches and seizures. This protection had formerly been construed primarily in cases involving intrusion into a physical place, but in *Katz* (at 351) the justices famously held that the Fourth Amendment "protects people, not places."

The moral interest at stake in data-protection regulation has seemed unclear to many. Turkington (1990), among others, has suggested identifying this interest as "informational privacy." Another complementary approach can be understood by returning to Clarke's (1994) notion of the "digital persona" that has increasingly become part of an individual's social identity. From this perspective, control over personal information is control over an aspect of the identity one projects to the world, and the right to privacy is the freedom from unreasonable constraints on the construction of one's own identity. This idea is appealing for several reasons: it goes well beyond the static conception of privacy as a right to seclusion or secrecy, it explains why people wish to control personal information, and it promises detailed guidance about what kinds of control they might wish to have.

Bellotti and Samarajiva develop this line of thinking further. Their point of departure is the sociologist Erving Goffman's finely detailed description of the methods by which people project their personae. Goffman (1957) argued that personal identity is not a static collection of attributes but a dynamic, relational process. People construct their identities, he suggested, through a negotiation of boundaries in which the parties reveal personal information selectively according to a tacit moral code that he called the "right and duty of partial display." Goffman developed this theory in settings (e.g. public places) where participants could see one another face to face, but it has obvious implications for technology-mediated interactions. In particular, to the extent that a technology shapes individuals' abilities to negotiate their identities, Goffman's theories have implications for that technology's design.

Bellotti and Samarajiva attempt to draw out these implications in different contexts, Bellotti in her experiments with media spaces that interconnect users' workspaces with video and data links and Samarajiva in his study of an online platform being developed in Quebec. These authors explore the conditions under which individuals can exert control and receive feedback over the release of personal information, Bellotti emphasizing technical conditions and Samarajiva emphasizing institutional conditions. Both authors describe conditions under which pathological relationships might arise. Goffman's theory, by helping articulate the nature of the wrongs in such cases, also helps specify how technology might help us avoid them. Technology cannot, of course, guarantee fairness in human relationships, but it can create conditions under which fairness is at least possible; it can also undermine the conditions of fairness.

CNID (also known, somewhat misleadingly, as Caller ID) illustrates the point. CNID is a mechanism by which a switching system can transmit a caller's telephone number to the telephone being called. The recipient's telephone might display the number or use it to index a database. CNID seems to raise conflicting privacy interests: the caller's right to avoid disclosing personal information (her telephone number) and the recipient's right to reduce unwanted intrusions by declining to answer calls from certain numbers. To reconcile these interests, CNID systems gener-

Table 1

	Subject-subject interaction	Subject-institution interaction
Negotiation of personal boundary	Bellotti (media spaces)	Samarajiva (online platform)
Binary access	Phillips (cryptography)	Burkert (personal identity)

ally come equipped with "blocking" options. Callers who do not wish to transmit their telephone number can block CNID; recipients may decline to answer calls that are not accompanied by CNID information. In effect, the negotiation of personal identity that once began with conventional telephone greetings ("Hello?"; "Hello, this is Carey; is Antonia there?"; "Oh, hi Carey, it's Antonia; how are you?") now begins with the decision whether to provide CNID information and the decision whether to answer the phone. Sharp conflict, however, arises in regard to the details of the blocking interface. If CNID is blocked by default then most subscribers may never turn it on, thus lessening the value of CNID capture systems to marketing organizations; if CNID is unblocked by default and the blocking option is inconvenient or little-known, callers' privacy may not be adequately protected. In 1995 these considerations motivated a remarkable campaign to inform telephone subscribers in California of their CNID options. The themes of default settings and user education recur in Bellotti's study, in which one system failed to achieve a critical mass of users because so many users had unintentionally kept it turned off.

It is useful to compare this approach to privacy with the approach taken by many proponents of PETs. Representative studies from this volume might be arranged in two dimensions, as shown in table 1. The horizontal axis distinguishes cases according to the structure of the interaction—between individual subjects or between a subject and an institution such as a bank. This distinction is obviously artificial, since it excludes intermediate cases, additional relevant parties, and other elements of context. The vertical axis pertains to the conceptualization of privacy, as the negotiation of personal boundaries or as the regulated and conditional access of some authority to sensitive personal information. (A third axis might contrast the normative standpoint of the designer or policymaker—see Bellotti and Burkert—with the empirical standpoint of the sociologist—

see Phillips and Samarajiva.) In Phillips's study of public contests over cryptography, the relevant authority is the government and the sensitive information is the cryptographic keys that can reveal the content of personal communications; in Burkert's analysis of technologies for conducting anonymous transactions, the relevant authorities are numerous and the sensitive information is personal identity. These authors' approach is labeled "binary access" because the issue at stake is whether the authority can gain access to the information; this access might be conditional (for example, upon a court order), but once granted it opens up a whole realm of personal information, often covertly, to a broad range of unwelcome uses.

The two rows in table 1 contrast in instructive ways. Personal boundary negotiation emphasizes personal choice, reciprocity, and fine-grained control over information in the construction of personal identity. Binary access emphasizes individual rights, the defense of interests, and coarse-grained control over information in the protection of personal identity. This distinction partly reflects real differences in the social relationships that two rows describe, but for some authors it also reflects different underlying approaches to privacy problems. Burkert points out that PETs are not inherently confined to the maintenance of anonymity; they could be used as tools for the finer-grained negotiation of identity. Mayer-Schönberger associates technologies for local choice with erosion of the values of social democracy and emphasizes that individuals have found themselves confused and overwhelmed by a proliferation of mechanisms that require them to make choices about esoteric matters that are more suited to technical analysis and legislation. Other authors, undaunted by this potential, have embraced technologies for local choice as expressions of market freedom. In each case, the actual conditions of fairness and the effective protection of interests in concrete situations are largely unknown.

The new technologies also have implications for conceptions of relationship, trust, and public space. Samarajiva observes that technology and codes of practice determine whether data-based "relationships" between organizations and individuals are fair, or whether they provoke anxiety. These concerns, which traditionally motivate data-protection regulation, are amplified by technologies that permit organizations to maintain highly customized "relationships" by projecting different organizational personae to different individuals. Such "relationships" easily become asymmetric, with the organization having the greater power to control what information about itself is released while simultaneously obscuring the nature and scope of the information it has obtained about individuals. Phillips describes a controversy over the conditions under which individuals can establish private zones that restrict access by outsiders. A secure telephone line is arguably a precondition for the establishment of an intimate relationship, an interest long regarded as a defining feature of human dignity (see, e.g., Inness 1992). This concern with the boundaries that are established around a relationship complements Bellotti's concern with the boundaries that are negotiated within a relationship. It also draws attention to the contested nature of those boundaries.

Beneficial relationships are generally held to require trust. As the information infrastructure supports relationships in more complex ways, it also creates the conditions for the construction of trust. Trust has an obvious moral significance, and it is economically significant when sustained business relationships cannot be reduced to periodic zero-sum exchange or specified in advance by contract. Phillips and Burkert both emphasize the connection between trust and uncertainty, but they evaluate it differently. For Phillips, trust and uncertainty are complementary; cryptography establishes the boundaries of trust by keeping secrets. Burkert, however, is concerned that this approach reduces trustworthiness to simple reliability, thereby introducing tacit norms against trusting behavior. Just as technology provides the conditions for negotiating relationships, it also provides the conditions for creating trust. Samarajiva points to the institutional conditions by which a technical architecture comes to support these conditions or else evolves toward a regime of coercive surveillance.

Public spaces have traditionally been understood as primary sites for the conduct of politically significant activities, and systematic surveillance of those spaces may threaten the freedom of association. Davies describes the shifting ways in which public discourse in the United Kingdom since the 1980s has constructed the issue of surveillance in public spaces. The introduction of inexpensive video cameras has brought a tension between privacy and personal security, generally to the detriment of the political value of privacy. Bellotti points out that the new technologies are being used in ways that erode the distinction between public and private space and in ways that problematize the very idea of private space by establishing long-lived interconnections among formerly separate spaces. Although occasioned by technological innovations, these observations converge with recent attempts to renegotiate the concepts of public and private and the conceptions of intimate relationship and political discourse that have traditionally gone with them.

Taken together, these considerations describe a complex new terrain around the central issues of voluntariness and coercion. Davies observes that organizations often identify disclosures of personal information as voluntary, even when the consequences of disclosure are unclear, when alternative courses of action are unavailable, or when failures to disclose are accompanied by unreasonable costs. This contested terrain was first mapped by the data-protection model, with its concepts of notification and transparency. The purpose of these concepts was to enable individuals to bargain more effectively over the disclosure and use of personal information. The emerging technological options, however, create a more complicated range of alternatives for policy and design. A fundamental division of labor emerges: some decisions about privacy are made collectively at the level of system architecture, and others are made individually through local bargaining. System architecture necessarily embodies social choices about privacy; these choices can make abuses more difficult, but they can also prevent individuals from tailoring their technology-mediated relationships to their particular needs. System architecture, however, rarely suffices to shape social outcomes, and architectural choices must always be complemented by policy measures such as regulations and sectoral codes. The data-protection model analyzed the relationship between architecture and policy in a simple, powerful way. Now, however, other possible analyses are becoming perceptible on the horizon.

Economic and Technical Scenarios

Privacy issues, then, pertain to the mechanisms through which people define themselves and conduct their relationships with one another. These mechanisms comprise technologies, customs, and laws, and their reliability and fairness are necessary conditions of a just social order. Those who raise concerns about privacy propose, in effect, to challenge the workings of institutions. Disputes about privacy are, among other things, contests to influence the historical evolution of these institutions. Before considering these contests in detail, though, let us consider the economic and technical logics that, some have held, are truly driving the changes that are now under way. My purpose is not to present these scenarios as adequate theories of the phenomena, but rather to make them available for critical examination.

A useful point of departure is Casson's (1994) analysis of the role of business information in the evolution of social institutions. Casson observes that information and markets have a reciprocal relationship: perfect markets require perfect information, but information is usually not free. To the contrary, information is one more good that is traded in the market. As information becomes cheaper (for example, through new technology), transaction costs are reduced accordingly and the conditions for perfect markets are better approximated (Coase 1937; Williamson 1975). This theory makes the remarkable prediction that many social institutions, having originated to economize on information costs, will break down, to be replaced by institutions that more closely resemble the market ideal of individually negotiated transactions. Advance ticket sales through networked booking systems, to take a simple example, now supplement box-office queues as a mechanism for allocating movie tickets. On a larger scale, the gradual breakdown of fixed social roles and of the customary responsibilities that go with them is facilitated by technological changes, particularly in communication and in record keeping, that make it easier to establish and evaluate individual reputations.

This theory suggests two consequences for privacy. The first is that individualized transactions must be monitored more closely than custombound transactions. By way of illustration, Casson offers a science fiction story about the future of road transportation in a world of very low information costs. Nowadays most roads are provided collectively, and their use is governed by customary mechanisms (such as traffic lights and right-of-way rules) that permit drivers to move toward their destinations without colliding very often. From an economic standpoint, this scheme has obvious inefficiencies: road utilization is uneven, congestion is common, and drivers spend much time waiting. With lower information costs, however, roads could operate more like railroads. Drivers wishing to go from point A to point B would call up a reservation system and bid for available itineraries, each precisely specifying the places and times of driving. Drivers' movements would be still regulated by traffic signals, but the purpose of the signals would now be to keep the drivers within the space-time bounds of the journey they had purchased. Lower-paying drivers would be assigned slower and less scenic routes, other things being equal, than higher-paying drivers. Overall efficiency would be maximized by the market mechanisms embodied in the reservation system. Issues of technical workability aside, Casson points out that such a scheme would raise concerns through its reliance on detailed monitoring of drivers' movements. Nor are these concerns entirely hypothetical, in view of the automatic toll-collection technologies now being deployed as part of the Intelligent Transportation Systems program (Agre 1995). Decreasing information costs make toll roads cheaper to operate, thus contributing to their spread. Information costs alone, of course, do not explain the full political and institutional dynamics of these developments, but they do lower one barrier to them.

The second consequence of Casson's theory relates to the evolution of privacy regulation itself. Data-protection regulation, on this theory, is essentially a set of customary responsibilities imposed on organizations that gather personal information. The theory further suggests that these responsibilities are economically inefficient, inasmuch as they preclude individualized negotiations between organizations and individuals about the handling of each individual's information. The "one size fits all" regulations are efficient, however, if the costs of individualized negotiation are high. Data-protection regulation is thus similar to the phenomenon of standardized contracts, which also economize on transaction costs, and regulation (as opposed to market competition between different sets of standardized contract terms) is required because the parties to a standardized contract hold asymmetrically incomplete information about the real costs and benefits of various information-handling policies. New information technologies reopen these questions in two ways: by making it economically feasible in some cases for organizations to negotiate the handling of customers' information on a more individualized basis, and by providing policymakers with a broader range of possible information-handling policies to impose on organizations that transact business with the public. Casson's analysis focuses on the first of these points, predicting a transition from generalized regulation to localized negotiation of privacy matters.

These two implications of Casson's theory may seem contradictory: more individualized market transactions require greater monitoring and therefore less privacy, and decreases in transaction costs permit a transition toward local negotiation and thus more efficient allocation of rights to personal information. Both of these contradictory movements are found in reality, however, and both movements are likely to continue. An economic optimist would suggest that the necessary outcome is a high level of *potential* monitoring whose *actual* level is regulated by the same allocative mechanisms that regulate everything else in the market. This scenario, however, makes numerous assumptions. It assumes, for example, that market forces will cause economic institutions to become perfectly transparent. This seems unlikely. After all, as Casson points out, information about information is inherently expensive because it is hard to evaluate the quality of information without consuming it. Economic efficiency can even be reduced by advertising and public-relations practices that frustrate consumers' attempts to distinguish between valid and bogus reputations. (Davies's chapter bears on this point in reference to the public construction of privacy issues by interested parties.) Most important, the optimistic economic scenario applies only in the longest possible term, after all transaction costs have been reduced to zero and the whole technical infrastructure of society has been revised to implement the efficient economic regime that results. It says little about the fairness of the constantly shifting and inevitably contested series of institutional arrangements that will govern social life between now and then.

Even if it is accepted, then, Casson's theory predicts only incremental shifts and should not be interpreted as arguing for a wholesale abandonment of regulation. Indeed, to the extent that transaction costs remain high, the theory argues that traditional protections should be retained. Advances in technology do not necessarily reduce transaction costs, and many such costs may have no relationship to technology. In some cases, as Rotenberg (1996) has pointed out, reductions in transaction costs should actually cause traditional protections to be strengthened. Posner (1981: 256, 264), for example, argues that high transaction costs should relieve magazine publishers and the Bureau of the Census of the obligation to obtain individuals' permission before making certain secondary uses of information about them; as technology lowers transaction costs, this argument is steadily weakened.

Assuming, however, that something approximating the economic scenario comes true, how might the necessary technologies of localized negotiation be implemented? Data-protection regulation, as we have seen, was originally motivated by fears about a single centralized government database, and it was subsequently forced to adjust its imagination to accommodate a world of wildly fragmented databases in both the public and the private sector. Immense incentives exist to merge these various databases, however, and many have predicted that the spread of computer networking will finally bring about the original dystopian vision. This vision has not come about on a large scale because of the great difficulty of maintaining the databases that already exist and because of the equally great difficulty of reconciling databases that were created using incompatible data models or different data-collection procedures (Brackett 1994).

It will always be difficult to reconcile existing databases. But newly created databases may be another story. Whereas the old dystopian scenario focused its attention on a single centralized computer, a new scenario focuses on standardized data models—data structures and identification and categorization schemes that emerge as standards, either globally or sectorally. Standards have contributed to the rise of institutions because they permit reliable knowledge of distant circumstances (Bowker 1994; Porter 1994). Many people once thought that it was impossible, for example, to categorize and intermingle grain from different farms (Cronon 1991). Once this was finally achieved, after several false starts, it became possible to trade commodities in Chicago and New York. Those cities became "centers of calculation" (Latour 1987) that gathered and used information from geographically dispersed sources. Information from multiple locations was commensurable, to within some controllable degree of uncertainty, because of the standardization of categories and measurement schemes. As a result, those with access to this combined information possessed a "view from nowhere" (Porter 1994) that permitted them to enter a tightly bound relationship with numerous places they had never visited.

Standardized data models may produce a similar effect, and for precisely the same reasons: standardized information is more valuable when it crosses organizational boundaries, and standardized real-world circumstances are more easily administered from a distance. A firm that plans to exchange information with other organizations can facilitate those transactions by adhering to standards, and a firm that expects to consolidate with other firms in its industry can increase its own value by standardizing its information assets.

It now becomes possible to employ distributed object database technology (Bertino and Ozsu 1994) to share information in an extremely flexible fashion. To be truly dystopian, let us imagine a single, globally distributed object database. Every entity instance in the world, such as a human being, would be represented by a single object, and all information about that person would be stored as attributes of this object. Individual organizations would still retain control over their proprietary information, but they would do so using cryptographic security mechanisms. At any given time, each organization would have access to a certain subspace of the whole vast data universe, and it would be able to search the visible portion of that data space with a single query. Organizations wishing to make certain parts of their database public (for example, firms wishing to make product information available to customers) would lift all security restrictions on the relevant attributes. Organizations with data-sharing agreements would selectively provide one another with keys to the relevant attributes in their respective segments of the data space. Real-time markets in data access would immediately arise. Lowered transaction costs would permit access to individual data items to be priced and sold. Data providers would compete on data quality, response time, and various contract terms (such as restrictions on reuse). Once the global database was well established, it would exhibit network externalities, and whole categories of organizations would benefit from joining it, for the same reasons that they benefit from joining interoperable networks such as the Internet. (Database researchers such as Sciore, Siegel, and Rosenthal (1994) speak of "semantic interoperability," but they are usually referring to systems that render existing heterogeneous databases interoperable by translating each database's terms in ways that render them commensurable with one another.)

Such a database would, of course, be extraordinarily difficult to get started. Bowker and Star (1994) illustrate some of the reasons in their study of an ambitious global project to standardize disease classifications. But the standardization of grain was difficult too. Standardization is a material process, not just a conceptual exercise, and standardized databases cannot emerge until the practices of identification and capture are standardized. The point is that considerable incentives exist to perform this work. On this view, privacy is not simply a matter of control over data; it also pertains to the regimentation of diverse aspects of everyday life through the sociotechnical mechanisms by which data are produced (Agre 1994).

Constructing Technology and Policy

Although powerful as stimuli to the imagination, the scenarios in the previous section are too coarse to account for the social contests through which privacy issues evolve. The authors in this volume paint a coherent and powerful picture of these contests—a political economy of privacy that complements and extends the work of Gandy (1993). This section summarizes their contributions, focusing on the interactions among technology, economics, and policy. This theory starts with observations about the somewhat separable logics of these three domains.

technological logic   Among the many factors influencing the evolution of technical systems, some are internal to the engineering disciplines that design them. Bijker (1987) refers to the "technological frames" that shape engineers' understandings of the problems they face and subsequently shape outsiders' understandings of the technologies themselves. Agre describes some of the metaphors—processing and mirroring—that help define a technological frame for the design of a computer system. Another aspect of this frame is the historically distant relationship between the designers of a system and the people whose lives the system's data structures represent. Although system designers' technological frame has changed slowly over the history of the computer, Bennett observes that the technology has improved rapidly in speed and connectivity. As a result, the underlying representational project of computing—creating data structures that mirror the whole world—has found ever-more-sophisticated means of expression in actual practices. Other technological systems—for example, the infrastructures of transportation, communication, and finance—embed their own disciplinary logic that shapes the privacy issues that arise within them.

economic logic   Privacy issues have evolved in the context of several trends in the global economy. Samarajiva points to the decline of the mass market and the proliferation of "compacks"—packages of products and services that are, to one degree or another, adapted to the needs of increasingly segmented markets and even particular customers. The flexible production of compacks presupposes a decrease in the costs of coordinating dispersed manufacturing and service activities, and the marketing of compacks presupposes a decrease in the costs of tracking the market and maintaining tailored relationships with numerous customers. Samarajiva also points to the significance of network externalities (which strongly condition attempts to establish new information and communication networks) and the peculiar economics of information (which can be dramatically less expensive to distribute than to produce). These effects imply that classical economic models may be a poor guide to the success of new social institutions.

policy logic   The authors in this volume are particularly concerned with the dynamics of the policy process. Moving beyond a normative consideration of privacy policies, they reconstruct the evolution and the implementation of these policies. Their central claims have already been sketched. Bennett observes that privacy policy has emerged from a global network of scholars—an "epistemic community" (Drake and Nicolaïdis 1992) whose thinking develops in a coordinated fashion. Conflicts arise, as might be expected, through organizations' attempts to quiet privacy concern in the public sphere (Davies) and the countervailing initiatives of other policy communities (Phillips). Privacy issues are distinctive, though, in their rapid globalization and in the oft-remarked mismatch between their high level of abstract concern and their low level of concrete mobilization.

Expressed in this way, these points seem to hover outside history. Yet they are necessary preliminaries to consideration of the numerous modes of interaction among the respective logics. Perhaps the most striking aspect of this interaction is the recurring sense, remarked by Flaherty and Davies, that privacy is a residual category—something left over after other issues have staked their claims. Privacy often emerges as a "barrier" to a powerful technical and economic logic, and privacy policy, in its actual implementation, often seems to ratify and rationalize new technical and institutional mechanisms rather than derailing them or significantly influencing their character. Part of the problem, of course, is a mismatch of political power. But it is also a problem of imagination: when privacy concerns arise in response to particular proposals, they can appear as manifestations of a perverse desire to resist trends that seem utterly inevitable to their proponents and to broader institutional constituencies.

With a view to understanding this phenomenon, and at the risk of some repetition, let me sketch some of the reciprocal influences among the technological, economic, and policy logics of privacy.

technological → economic   Technology influences economic institutions in several ways. Perhaps most obvious, technology largely determines the practicalities of production, and classical economics holds technical factors constant when exploring the conditions of market equilibrium. In a more subtle way, technology serves a rule-setting function. The computers and networks of a stock trading system, for example, embody and enforce the rules of trading. The rule-setting function is also familiar from the previous section's analysis of technology's role in setting ground rules for the negotiation of individual relationships. Finally, technology affects economic institutions through its effect on information costs and other transaction costs. As Casson points out, these effects can be far-reaching and qualitative and can provide a major impetus for the economic monitoring that raises privacy concerns.

technological → policy   Technology influences policy formation, Phillips observes, through the obduracy of its artifacts: once implemented, they are hard to change. Technological practices are obdurate as well; PETs represent a rare occasion on which technical practitioners have revealed fault lines within an existing body of technical practices, so that formerly inevitable technical choices begin to admit alternatives. Technology thus influences policy in a second way: by determining the range of technological options available for addressing public concerns. Flaherty's tale of medical prescription systems represents a rare occasion on which a data-protection commissioner has been able to affect the architecture of a proposed system, but at least the option was technically available. The dynamics of privacy policy, as Bennett points out, have also been shaped by the global nature of computer networking, which is even more indifferent to jurisdictional boundaries than the telephone system. Technology also influences policy, finally, by providing the infrastructure for contesting policy, as in the case of privacy activists' use of the Internet.

economic → technological   Technology acquires its obduracy, in part, through the economic dynamics of standards. Once established in the market as the basis for compatibility among large numbers of otherwise uncoordinated actors, standards tend to be reproduced regardless of their consequences for concerns such as privacy, or even regardless of their efficiency in a changed technological environment (David 1985). The standards themselves, meanwhile, arise through practices of computer system design that have been shaped across generations through the demands of business applications. The full extent of this influence is not generally visible to the practitioners, who inherit a seemingly rational body of practices and who encounter "problems" in the bounded, conventionalized forms that these practices define. Yet the influence becomes visible suddenly as technical practices are moved from applications that structure relationships within the scope of a business to applications that structure relationships with other parties. A current example: Client-server systems arose on the premise that users do not wish do "see" the boundary between the client and the server. This is a reasonable commitment when that boundary has no legal or moral significance. It is not at all reasonable, however, for consumer applications—such as commerce on the Internet—in which that boundary corresponds precisely to the sphere over which individual users wish to maintain informational control. In such cases, it suddenly becomes important for the boundary between client and server to become visible, and the invisibility of this boundary becomes the raw material for rumors and scams.

economic → policy   Flaherty notes several possible effects of economic phenomena on the policy process. Perceptions of the overall health of the economy may, other things being equal, influence the attention devoted to privacy issues. The organization of an industry affects its capacity to mobilize politically to define issues and shape regulatory regimes. Budgetary pressures on government agencies create organizational incentives for privacy-threatening initiatives that might not otherwise have materialized. Other economic phenomena affect the substance of policy issues and the institutional realities with which any policy will interact in its implementation. For example, Bennett points out that the economic properties of personal information have contributed to the global nature of privacy problems and to the global harmonization of privacy policy. Samarajiva's chapter is a concrete study in the complex arrangement of economic incentives that motivated one enterprise to establish relatively strong privacy policies. Because the system was going to be viable only if 80 percent of the potential members of the audience took specific actions to sign up, it became necessary to appease privacy fundamentalists by providing customers with a high degree of transparency and control over personal information. These policies were largely private rather than public, but they were congruent with the strong regulatory regime in Quebec. Similar considerations may explain the willingness of industries to submit to particular regulatory systems.

policy → economic   Little work has been done to evaluate the economic impact of privacy policy. Reporting requirements impose costs, and rules about the secondary use of personal information affect business models and may sometimes determine their viability. On the other hand, when organizations review their information-handing practices with a view to implementing new privacy policies, they frequently discover opportunities for improvements in other areas. Privacy-protection measures may also reduce the economic costs associated with identity theft and other fraudulent uses of personal information. Policy may also influence economic institutions by shaping technology, though any massive effects of this type still lie in the future.

policy → technological   Historically, privacy policy has not attempted to influence the basic architecture of information systems. Burkert and Agre, however, point to a joint report of Ontario's Information and Privacy Commissioner and the Netherlands' Registratiekamer (1995) that ex-

Encouraging the adoption of such technologies, unfortunately, requires more than technical existence proofs. One significant issue is trust in the system. After all, the system could also be used as a traditional biometric authentication system through the encoding of a universal identifier; an organization that simply claims to be using biometric encryption is therefore not necessarily protecting anybody's privacy. The issue has already arisen in Ontario, where a social activist group, the Guelph Coalition Against the Cuts (1996), has begun a campaign against the provincial government's plan to employ biometric encryption in distributing welfare payments. The Coalition suggests that this "technical proposal for treating the poor like criminals" is a suitable symbol of the conservative government's cuts in social-welfare programs. Fingerscanning systems for welfare recipients have, after all, been implemented in the United States; the issue here is whether the Ontario system based on biometric encryption deserves to be assimilated to the police-state metaphors that have long shaped the public's understanding of universal identification. It is conceivable that any misunderstandings could be repaired through a suitable public communication campaign. One potential model might be the successful campaign that Pacific Bell conducted in the spring of 1996 as a condition of its introduction of Caller ID service to California. Pseudoidentity, unfortunately, is a more technical concept than Caller ID, and the technology itself is counterintuitive even to computer professionals.

Indeed, the force of habit among system designers is another significant obstacle to the adoption of PETs. So long as the language of computer system design tends to blur the difference between representations and reality, treating data records as mirror images of the world, the protection that PETs provide for individual identity will remain nearly incomprehensible—designers will often simply assume that records pertaining to individuals can be traced back to them. The problem does not lie in the practices of database design, which may evolve toward the use of identifers that can equally easily be universal or pseudonymous—in that limited sense, the methods of computer system design do not, as social theorists have often assumed, inherently lead to the invasion of privacy. Instead, privacy invasion results from the way in which the technical methods are customarily applied. The necessary change in customs can be encouraged through model programs and exhortation from policymakers.

Ultimately, though, it will be necessary to revise the training of system designers to integrate the new range of technical options. Lessons will be needed in cryptography and protocols for the management of pseudonyms, of course, but other necessary revisions are less obvious. The examples provided in database texts, for example, almost invariably identify human beings by their names, and they rarely provide any sense of the moral issues that are at stake in the selection of primary keys.

Much experience will have to accumulate before PETs can be integrated into the wide variety of institutions and concrete life situations to which they are applicable in theory (see Bijker and Law 1992). Troubles arise immediately when organizational relationships are not confined to digital media. Customer-service telephone numbers, for example, have typically required individuals to identify themselves, and telephone interactions based on pseudonyms will inevitably be clumsy. Customers who enjoy the record-keeping benefits of periodic statements (from credit cards, banks, the phone company, and now automatic toll-collection systems) will require some way of obtaining such statements without disclosing their names and mailing addresses.

Perhaps the most significant accomplishment of PETs to date is to have initiated a shift of imagination. When records of personal information can always be traced back to the individuals whose lives they represent, privacy interests trade off against system functionality. The result, as the historical record plainly shows, is that privacy interests almost invariably lose out in the end. Data-protection regulation can contain the abuse of personal information within this framework, but it cannot contain the proliferation of technologies that create and use personal information for ever-broader uses. Markets in privacy can operate at the boundaries of the system (for example, to express customer choice about secondary uses of transaction-generated information), but such markets require that customers be able to estimate the disutility of secondary use despite the enormous credit-assignment problem. (The issue is one of transparency: "How did they get my name?") PETs, however, promise to provide data subjects with much more detailed control over the use of their information, and to greatly lower the price of refusing to disclose it. Whether that promise is fulfilled, however, depends on a great diversity of factors.

Conclusion

I hope to have suggested the utility of historical investigation of computing as a representational practice. A brief chapter cannot survey the whole field of computer science or even the whole subdiscipline of systems analysis; numerous episodes in the complex history of data have been glossed over, and relevant areas such as data communications and security have been left out entirely. Nonetheless, some clear trends can be discerned. At stake is the sense in which a technical field has a history: what it inherits from past practice, how this inheritance is handed down, the collective thinking through which the field evolves, and how all this is shaped by the institutional contexts within which the work occurs.

To answer these questions, one might suggest that privacy problems have arisen through the application of industrial methods to nonindustrial spheres of life, where normative relations of representation and control are different. But that statement alone is too simple. Technical work is not indifferent to the contexts in which it is applied; quite the contrary, it is continually confronted by the practical problems that arise in pursuing instrumental goals in particular concrete settings. Another part of the problem, then, is the manner in which "problems" arise in the course of practitioners' work, and how these problems are understood. Database designers, for example, have been forced to clarify their methods on numerous occasions when existing databases have been used for new and unforeseen purposes.

Yet these mechanisms of practical feedback were evidently not adequate to stimulate the invention or the ready adoption of methods to decouple data records from human identity. The mathematicians who did invent such methods in the 1980s were explicitly motivated by a desire to protect privacy, and they have faced an uphill fight in getting their ideas adopted. Much of this fight, of course, has been overtly political (see Phillips's chapter in this book). Organizations with pecuniary interests in the secondary use of personal information will presumably not be enthusiastic about the new technologies. On another level, however, the fight concerns basic understandings of representation and its place in human life. Information is not an industrial material or a mirror of a pre-given reality. It is, quite the contrary, something deeply bound up with the material practices by which people organize their lives. Computer systems design will not escape its association with social control until it cultivates an awareness of these material practices and the values they embody.

Acknowledgements

One version of this paper was presented at the University of Toronto, where I benefited from the comments of Andrew Clement. Another version was presented at AT&T Labs, where Paul Resnick, Joel Reidenberg, and several others offered helpful comments. Geoff Bowker, Michael Chui, Roger Clarke, and Leigh Star kindly offered suggestions on a draft.

References

Agre, Philip E. 1994. Surveillance and capture: Two models of privacy. *Information Society* 10, no. 2: 101–127.

Agre, Philip E. 1995a. Institutional circuitry: Thinking about the forms and uses of information. *Information Technology and Libraries* 14, no. 4: 225–230.

Agre, Philip E. 1995b. From high tech to human tech: Empowerment, measurement and social studies of computing. *Computer Supported Cooperative Work* 3, no. 2: 167–195.

Agre, Philip E. 1997. *Computation and Human Experience*. Cambridge University Press.

Benedikt, Michael, ed. 1991. *Cyberspace: First Steps*. MIT Press.

Beniger, James R. 1986. *The Control Revolution: Technological and Economic Origins of the Information Society*. Harvard University Press.

Bennett, Colin J. 1992. *Regulating Privacy: Data Protection and Public Policy in Europe and the United States*. Cornell University Press.

Berg, Marc. 1997. *Rationalizing Medical Work: Decision-Support Techniques and Medical Practices*. MIT Press.

Bijker, Wiebe E., and John Law, eds. 1992. *Shaping Technology / Building Society: Studies in Sociotechnical Change*. MIT Press.

Borgida, Alexander. 1991. Knowledge representation, semantic modeling: Similarities and differences. In *Entity-Relationship Approach: The Core of Conceptual Modeling*, ed. H. Kangassalo. North-Holland.

Bowers, C. A. 1988. *The Cultural Dimensions of Educational Computing: Understanding the Non-Neutrality of Technology*. Teachers College Press.

Bowers, John. 1992. The politics of formalism. In *Contexts of Computer-Mediated Communication*, ed. M. Lea. Harvester Wheatsheaf.

Brachman, Ronald J. 1979. On the epistemological status of semantic networks. In *Associative Networks: Representation and Use of Knowledge by Computers*, ed. N. Findler. Academic Press.

Braverman, Harry. 1974. *Labor and Monopoly Capital: The Degradation of Work in the Twentieth Century*. Monthly Review Press.

Bud-Frierman, Lisa. 1994. *Information Acumen: The Understanding and Use of Knowledge in Modern Business*. Routledge.

Chaum, David. 1985. Security without identification: Transaction systems to make Big Brother obsolete. *Communications of the ACM* 28, no. 10: 1030–1044.

Chaum, David. 1986. Demonstrating that a public predicate can be satisfied without revealing any information about how. In *Advances in Cryptology—CRYPTO '86*, ed. A. Odlyzko. Springer-Verlag.

Chaum, David. 1990. Showing credentials without identification: Transferring signatures between unconditionally unlinkable pseudonyms. In *Advances in Cryptology—AUSCRYPT ' 90*, ed. J. Seberry and J. Pieprzyk. Springer-Verlag.

Chaum, David. 1992. Achieving electronic privacy. *Scientific American* 267, no. 2: 96–101.

Chaum, David Amos Fiat, and Moni Naor. 1988. Untraceable electronic cash. In *Advances in Cryptology—CRYPTO '88*, ed. S. Goldwasser. Springer-Verlag.

Chen, Peter Pin-Shan. 1976. The entity-relationship model: Toward a unified view of data. *ACM Transactions on Database Systems* 1, no. 1: 9–37.

Clarke, Roger. 1994. Human identification in information systems: Management challenges and public policy issues. *Information Technology and People* 7, no. 4: 6–37.

Clement, Andrew. 1988. Office automation and the technical control of office workers. In *The Political Economy of Information*, ed. V. Mosco and J. Wasco. University of Wisconsin Press.

Couger, J. Daniel. 1973. Evolution of business system development techniques. *Computing Surveys* 5, no. 3: 167–198.

Derrida, Jacques. 1976. *Of Grammatology*. Johns Hopkins University Press. Originally published in French in 1967.

Di Genova, Frank, and S. Kingsley Macomber. 1994. Task Order Contract for Emissions Inventory Projects: Final Report. Report SR94-01-01, prepared for California Air Resources Board by Sierra Research.

Dourish, Paul. 1993. Culture and control in a media space. In *Proceedings of the Third European Conference on Computer-Supported Cooperative Work*, ed. G. de Michelis et al. Kluwer.

Edwards, Paul N. 1996. *The Closed World: Computers and the Politics of Discourse in Cold War America*. MIT Press.

Feenberg, Andrew 1995. *Alternative Modernity: The Technical Turn in Philosophy and Social Theory*. University of California Press.

Flaherty, David H. 1989. *Protecting Privacy in Surveillance Societies: The Federal Republic of Germany, Sweden, France, Canada, and the United States.* University of North Carolina Press.

Friedman, Andrew L. 1989. *Computer Systems Development: History, Organization and Implementation.* Wiley.

Friedman, Batya, and Helen Nissenbaum. 1994. Bias in Computer Systems. Report CSLI-94-188, Center for the Study of Language and Information, Stanford University.

Gelernter, David. 1991. *Mirror Worlds, or the Day Software Puts the Universe in a Shoebox.* Oxford University Press.

Gilbreth, Frank B. 1921. *Motion Study: A Method for Increasing the Efficiency of the Workman.* Van Nostrand.

Guelph Coalition Against the Cuts. 1996. Campaign against fingerscanning. Message circulated on Internet, July 10.

Guillén, Mauro F. 1994. *Models of Management: Work, Authority, and Organization in a Comparative Perspective.* University of Chicago Press.

Halpert, Julie Edelson. 1995. Driving smart: A new way to sniff out automobiles that pollute. *New York Times,* October 22.

Information and Privacy Commissioner (Ontario) and Registratiekamer (Netherlands). 1995. Privacy-Enhancing Technologies: The Path to Anonymity (two volumes). Information and Privacy Commissioner (Toronto) and Registratiekamer (Rijswijk).

Lyon, David. 1994. *The Electronic Eye: The Rise of Surveillance Society.* University of Minnesota Press.

March, James G., and Johan P. Olsen. 1989. *Rediscovering Institutions: The Organizational Basis of Politics.* Free Press.

Molina, Alfonso H. 1993. In search of insights into the generation of techno-economic trends: Micro- and macro-constituencies in the microprocessor industry. *Research Policy* 22, no. 5–6: 479–506.

Montgomery, David. 1987. *The Fall of the House of Labor: The Workplace, the State, and American Labor Activism, 1865–1925.* Cambridge University Press.

Nelson, Daniel. 1980. *Frederick W. Taylor and the Rise of Scientific Management.* University of Wisconsin Press.

Pool, Ithiel de Sola. 1983. *Technologies of Freedom.* Harvard University Press.

Radin, George. 1983. The 801 minicomputer. *IBM Journal of Research and Development* 27, no. 3: 237–246.

Registratiekamer. See Information and Privacy Commissioner.

Reingruber, Michael C., and William W. Gregory. 1994. *The Data Modeling Handbook: A Best-Practice Approach to Building Quality Data Models.* Wiley.

Rogers, Paul. 1996. Your future car may be a spy: Clean-air proposal raises privacy concerns. *San Jose Mercury News,* June 14.

Shaw, Anne G. 1952. *The Purpose and Practice of Motion Study*. Harlequin.

Simsion, Graeme C. 1994. *Data Modeling Essentials: Analysis, Design, and Innovation*. Van Nostrand Reinhold.

Star, Susan Leigh. 1995. The politics of formal representations: Wizards, gurus and organizational complexity. In *Ecologies of Knowledge*, ed. S. Star. SUNY Press.

Sterling, Bruce. 1993. War is virtual hell. *Wired* 1, no. 1: 46–51, 94–99.

Suchman, Lucy. 1992. Technologies of accountability: Of lizards and aeroplane. In *Technology in Working Order: Studies of Work, Interaction, and Technology*, ed. G. Button. Routledge.

Taylor, Frederick Winslow. 1923. *The Principles of Scientific Management*. Harper.

Thompson, Paul. 1989. *The Nature of Work: An Introduction to Debates on the Labour Process*, second edition. Macmillan.

Walkerdine, Valerie. 1988. *The Mastery of Reason: Cognitive Development and the Production of Rationality*. Routledge.

Yates, JoAnne. 1989. *Control through Communication: The Rise of System in American Management*. Johns Hopkins University Press.

2

Design for Privacy in Multimedia Computing and Communications Environments

Victoria Bellotti

Why is it acceptable for passersby to stare in through the window of a restaurant at the people dining there, but not through the window of a house to see what the occupants are up to? Why do people try on clothes in a fitting room, but not in other parts of a store? Why is it acceptable for you to read over the shoulder of a friend, but not over the shoulder of a stranger sitting in front of you on a bus?

In public and private places there are different more or less implicit rules about acceptable behaviors and interpersonal access rights. These rules depend upon the people involved and their roles and relationships, the places and their physical properties, and what those places are appropriated for. People learn these rules in the course of normal socialization, and (unlike children, who often amuse or embarrass us by breaking these rules) most adults adopt certain behaviors appropriate to the place they are in. For example, in Western society one tends to avoid walking unannounced into private places, and one behaves in particular ways in public. In public there are many activities which one would feel uncomfortable indulging in, such as licking one's plate or scratching one's behind, though most people wouldn't think twice about the same activities in their own home, even in front of the family. When we break the unwritten rules of private and public places, we become targets for disapproval and may be regarded as threatening or even insane (Goffman 1959). Complying with these rules requires that we have a good sense of the nature of the place we are in and the people who are in it.

In this chapter I address design issues relating to the implications of the blurring of boundaries between public and private places. Innovations in multimedia communications and computing technology increase the connections between places and the connections between people distributed in space, and as a result the intuitive sense of place and presence that governs our observable behavior can no longer be relied upon to ensure that we will not be seen, overheard, or even recorded. The physical space around us that we can readily perceive is not necessarily an indicator of the places or people that may be connected to it. A place that seems to be private may, in fact, be otherwise (Harrison and Dourish 1996).

My concern, then, is with how we enable people to determine that they are presenting themselves appropriately and with how to control intrusive access over computer-mediated communications infrastructures and other systems that connect distributed places and the people in them. I present a simple conceptual framework for design for privacy in computer-mediated communications and collaboration environments; I conclude with examples of its application.

Communications Growth and Privacy Erosion

In recent decades there has been steady growth in the use and manipulation of vast quantities and varieties of personal data supported by computing technology. While this is critical to government, public services, business, and the livelihood of many individuals, it can also facilitate unobtrusive access to, manipulation of, and presentation of what people might consider private data about themselves (Parker et al. 1990; Dunlop and Kling 1991), raising Orwellian specters of Big Brother in our minds as we anticipate the inevitable erosion of personal privacy. New multimedia communications and computing technology is potentially much more intrusive than traditional information technology because of its power to collect even more kinds of information about people, even when they are not directly aware that they are interacting with or being sensed by it.

There are two major classes of privacy concerns relating to communications and computing technology. The first class covers technical aspects of such systems, including the capabilities they support. Some may be designed for what many would consider insidious or unethical uses (such as surveillance, often by legitimate or official organizations) (Clarke 1988; Samarajiva, in present volume). Cultural censure and the law generally discourage more extreme applications (although these forces trail behind the advances in sophistication of the technology). Technical security issues also belong in this class and relate to the fact that information systems, and particularly distributed systems, are vulnerable to unofficial, covert subversion (Lampson et al. 1981). Although it can be made extremely difficult to tamper with data in computing systems, protection mechanisms are seldom secure in practice (Mullender 1989). Consequently, there is a healthy body of ongoing research devoted to useful models and standards for the protection of software (Mullender 1989; Lampson et al. 1981; Chaum 1992; Bowyer 1996; Denning and Branstad 1996).

The second class of privacy concerns relates to a less well understood set of issues associated with multimedia computing and communications systems. These issues do not necessarily have to do with technical aspects of computer and communications systems; rather, they arise from the relationship between user-interface design and socially significant actions. Multimedia computing and communications systems are beginning to emerge as means of supporting distributed communication and collaboration, often with users moving from place to place. Such systems may use video, audio, infrared, or other new media to offer many new opportunities for capturing and processing data. The user interfaces to such systems, which may foster unethical use of the technology, are also much more conducive to inadvertent intrusions on privacy (Heath and Luff 1991; Fish et al. 1992).

The need to understand and protect personal privacy in sophisticated information systems is becoming even more critical as computing power moves out of the box on the desk into the world at large. We are entering the age of ubiquitous computing (Weiser 1991), an age in which our environment will come to contain computing technology in a variety of forms. Microphones, cameras, and signal receivers used in systems for wireless communication will, in combination with networked computing infrastructures, offer the potential to transmit and store information speech, video images, or signals from portable computing devices, active badges (Want et al. 1992), electronic whiteboards (Pederson et al. 1993), and so on.

Multimedia information can be also be processed, accessed and distributed in a variety of ways. A variety of prototype systems now offer activity-based information retrieval, diary services, document tracking, audio-video (AV) interconnections, video-based awareness, and capture and replay of multiple streams of recorded work activity (Lamming and Newman 1991; Eldridge et al. 1992; Gaver et al. 1992; Dourish and Bly 1992; Hindus and Schmandt 1992; Minneman et al. 1995).

Ubiquitous computing implies embedding technology unobtrusively within all manner of everyday objects so that they can potentially transmit and receive information to and from other objects. The aims are not only to reduce the visibility of the technology but also to empower its users with more flexible and portable applications to support the capture, communication, recall, organization, and reuse of diverse information. The irony is that the unobtrusiveness of such technology both obscures and contributes to its potential for supporting invasive applications, particularly as users may not even recognize when they are on line in such an environment. Designers must therefore consider carefully how services that capitalize on such powerful technology can be designed without compromising the privacy of their users.

Interactions between people that are mediated by technology are prone to both conscious and inadvertent intrusions on privacy. Such intrusions often result from inadequate feedback about what information one is broadcasting and an inability to control ones accessibility to others. This class of concerns is my focus in this chapter. My aim is to provide some guidance for designers of computer-mediated communication (CMC) systems and computer-supported collaborative work (CSCW) systems in how to support privacy for their users.

Defining Privacy

If we are in the business of designing communication and collaboration systems to support privacy, it is useful to have a working definition of what privacy might mean. Two types of definition are common; I refer to these as *normative* and *operational*.

A normative definition of privacy involves a notion that some aspects of a person's nature and activities are normally regarded as private and should not be revealed to anyone. Reiman (1995) offers such a definition, according to which privacy is the condition in which others are deprived of access to you and in order for there to be a right to privacy there must be some valid form specifying that some personal information about or experience of individuals should be kept out of other individuals' reach. Such norms may be legal. This would presumably mean for designers that people should be accessible by one another at certain times and definitely not accessible at other times. Furthermore, some information should be available to others and some should not.

Reiman's definition is not convincing, even from his own arguments. The value of privacy and the legislation designed to protect it vary widely, both culturally and legally, between countries (Milberg et al. 1995). For example, in the United Kingdom there has been serious resistance to the idea of carrying and routinely presenting a photo ID—resistance based on the view that this is an intrusion into people's privacy—whereas in France and in the United States this is more acceptable.

A normative definition of privacy is also problematic because many other personal and contextual factors come into play in deciding if privacy has been violated. These factors might include whether two people are related, where something happens (e.g., at home, in the workplace, or in public), and whether someone is aware of what is taking place or mentally competent to give permission. This variability tends to result in lengthy court cases to determine whether a legal violation of privacy has occurred (Privacy Protection Study Commission 1991), often with plaintiffs being surprised and dismayed to find that there is no legal protection against their complaint (McClurg 1995). If the law is so complex and unintuitive, then it seems that no team of designers, consultants, and users, or anyone else for that matter, is likely to be able to agree upon who should have access to whom or what and in which circumstances.

An operational definition of privacy refers to a capability rather than to a set of norms. This capability can be thought of as *access control* (Bellotti 1996). For example, Stone et al. (1983) include in their definition of privacy the ability to control information about oneself. Similarly, Samarajiva (present volume) refers to privacy as the control of outflow of information that may be of strategic or aesthetic value to the person and the control of inflow of information (including initiation of contact).[1] These operational definitions of privacy seem much more practical than normative definitions for designers of CSCW systems, since they allow for a person's need to make individual, cultural, and contextually contingent decisions as to how much information he or she desires to exchange and when. They also lend themselves readily to the definition of a requirement for the design of access-control mechanisms: users must be able to know (have feedback) about and control the consequences of their interactions with technology in terms of how visible and accessible they and their information are to others.

As computing systems increasingly support information sharing and communication, it is important for people to understand how they and their information are accessible, when, and to whom. They must also be able to control that access easily and intuitively. Without feedback and appropriate control, even secure systems used by well-meaning work groups are untrustworthy, since it is always unclear to the individual user who has legitimate access to what. Even with secure systems and within the law, many writers claim that privacy on line is currently invaded on a regular basis, sometimes inadvertently. One of the main reasons why this happens is that *people do not generally appreciate what the state of their personal information on line is with respect to access by others, and thus fail to take steps to control it,* sometimes with serious consequences (Forester and Morrison 1990; Weisband and Reinig 1995).

My concern in the rest of this chapter is to help to make privacy tractable as a central design issue in its own right for developers of potentially intrusive technologies. I present a framework for addressing the design of control and feedback of information captured by multimedia computing and communications environments. While control and feedback about information and interpersonal access are fundamental to privacy, the other side of the coin is that they are also key resources for successful communication and collaboration among people. It is therefore doubly important that they be considered in the design of networked and collaborative computing and communication systems.

In the following discussion I direct my attention toward the domain of networked audio-video communications infrastructures, drawing upon experiences with two particular systems. However, this design framework may also be related to the design of other CSCW, CMC, and ubiquitous computing systems, particularly those that involve device tracking, image processing, and on-line activity monitoring—in general, to any system that facilitates, for its users, the ability to interact with or information about others. I begin with two specific examples of potentially intrusive systems which were designed to provide resources for remote collaboration and which raised serious issues about how users' privacy was to be respected and supported. I then go on to outline the design framework and how it can be applied to improving these systems to support privacy.

Maintaining Privacy in Media Spaces: Feedback and Control in RAVE

Media spaces (Stults 1988; Dourish 1993; Harrison et al., in press) are a recent development in ubiquitous computing technology, involving audio, video, and computer networking. They have been the focus of considerable research and industrial interest in support for distributed collaborative work (Root 1988; Mantei et al. 1991; Gaver et al. 1992; Fish et al. 1992; Tang et al. 1994; Bellotti and Dourish, in press).

In a media space, cameras, monitors, microphones, and speakers are placed in an office to provide an AV node that is controllable from a workstation. Centralized switching devices for analog AV (or specially designed software for digital AV) allow one to make connections in order to communicate and work with others and to be aware of what is going on remotely without leaving one's desk. A media space often includes public nodes, with cameras (but not microphones) stationed and left on in public areas. These nodes are freely accessible for remote connections from all other users of this media space.

While media-space technology improves the accessibility of people to one another, some may feel that their privacy is compromised. The very ubiquity of such systems means that many of the concerns with existing workstation-based information systems are aggravated. A much wider variety of information can now be captured. People are much less likely to be off line (inaccessible) at any given moment. Further, the design of many of these systems is such that it may not be clear when one is off or on line and open to scrutiny (Mantei et al. 1991; Gaver 1992). People also express concern about their own intrusiveness when they try to make contact without being able to determine others' availability (Cool et al. 1992).

RAVE (the Ravenscroft Audio-Video Environment), at the Rank Xerox EuroPARC[2] research laboratory in Cambridge was one of several analog

AV media spaces set up in various research laboratories around the world. At EuroPARC, where a media-space node was installed in every office, privacy was a major concern of all who were involved. As the installation process was underway, one of the researchers took on the role of "user consultant," interviewing prospective users and distributing questionnaires asking them about their requirements of the system and their fears about its potential intrusiveness. Design was also informed by studies of how collaborative work is socially organized and how such technology impacts it (Smith et al. 1989; Heath and Luff 1991, 1992). A series of design meetings focused on how to provide RAVE's users with appropriate services for connection to one another and to public spaces while enabling them to preserve their own privacy as they saw fit. Concerns about privacy were accommodated by ensuring that users could decide how accessible they would be to others via the media space (Dourish 1991; Gaver et al. 1992; Dourish 1993).

Two important principles emerged in designing for privacy with respect to accessibility in RAVE (Gaver et al. 1992). These can be defined as follows:

feedback   Informing people when and what information about them is being captured and to whom the information is being made available.

control   Empowering people to stipulate what information they project and who can get hold of it.

Feedback depended on the type of RAVE connection being made. Three kinds of interpersonal connection were glance, v-phone call and office-share. Glance connections were one-way, video-only connections of a few seconds' duration. V-phone and office-share connections were longer two-way AV connections. For glances, audio feedback (Gaver 1991) alerted the user to the onset and the termination of a connection and could even announce who was making it. For the two-way office connection, reciprocity of the connection acted as feedback about its existence (if I see your office, you see mine); in the case of an attempted v-phone connection, an audio ringing signal was given and the caller's name was displayed on the workstation, whereupon the recipient could decide whether to accept or reject the connection. Office-shares, being very long term, did not require such a protocol.

Public areas had cameras, which could be accessed by a glance or a background connection which was indefinite, one-way, and video-only. Feedback about the presence of a camera in a public place was provided in the form of a video monitor beside the camera which displayed its view.

RAVE users could control who could connect to them and what kind of connections each person was allowed make. If they omitted to do so, automatic defaults were set to reject connections. User control via the workstation was supported by a special interface to the complex AV signal-switching and feedback mechanisms (Dourish 1991). These mechanisms defined the different types of connections possible between people, to different public areas, and to media services (e.g., video players). Interpersonal connection capability lists manipulated from this interface defined who was permitted to connect to whom, thus giving long-term static control over accessibility.

Providing distinct connection types with different kinds of feedback also enabled users to exercise discriminating, dynamic control over their accessibility, as in the v-phone call. (For a fuller description of these features see Dourish 1993.)

The design evolved over an extended period of time, together with a culture of trust and acceptable practices relating to its use. Users were allowed to customize, or to ignore the technology as they saw fit. Thus, privacy, although an initial concern, came to be something people did not worry about very much. Eventually some of the initial objectors to the technology became its greatest proponents as they discovered that the benefits of RAVE exceeded their initial expectations and as their concerns about potential intrusions were met with design refinements.

Representing Remote Presences in Apple's Virtual Café

The User Experience Research Group in Apple's Advanced Technologies Group (ATG) has been seeking to explore ways in which remote colleagues and friends can enjoy a sense of awareness of one another's activities and co-presence through multimedia computing technologies. In the light of some of the understandings gained from experience with the RAVE media space, an attempt was made to provide a video-based awareness server akin to the Portholes system, which was developed on top of RAVE (Dourish and Bly 1992). Portholes was a system which, every 5 minutes or so, collected small bit maps of frame-grabbed video images from office nodes and public nodes in EuroPARC's media space in the United Kingdom and the associated Xerox PARC laboratory's media space in California and displayed them on people's workstation screens. This provided its users with a sense of what was going on at both sites. One of the concerns about media spaces, and consequently about Portholes, was that, while users were able to gain a sense of what was going on in one other's offices and in public places, those in the public places had no sense of who was connected to the public nodes (unlike the reciprocal arrangements for office nodes).

At Apple, we designed an awareness service (a partial media space in some senses) that displayed frame-grabbed images taken from a camera connected to a server machine in the Apple coffee bar, Oh La La, on a web page which people could visit. Remote users of this Virtual Café were able to see how long the line for coffee was. The privacy concerns here, as we expected, came from the fact that the coffee shop's staff had to agree to being on camera all day and from the fact that most of the people waiting for coffee didn't even know there was a camera pointing at them. A preliminary survey also revealed that a small but significant proportion of the customers were not happy to discover that they were on camera.

This installation was not technologically innovative; there were already many sites where cameras in public places were sending images to web pages. The difference here was that we were very much concerned with addressing the privacy issue and were determined to provide people in the public place with some kind of feedback about the fact that there was a camera pointing at them, and furthermore that there were specific people connected to that camera. Unfortunately, we were using a digital camera that was even less conspicuous than the analog cameras used in the RAVE media space. We put signs up, but people tended not to see them. We also asked remote users to type their names in an authentication page at the web site, and this information was downloaded to the awareness server in the café. However, with our first prototype, only the staff in the café had access to the names of the remote visitors displayed on the computer screen.

Disembodiment and Dissociation

A number of concerns with media spaces relate to the fact that cameras and microphones are unobtrusive and thus hard to detect. Further, in a media space they tend to be left on continuously, with extended connections to other places. It is also generally difficult to tell if they are switched on and impossible to see the connections being made to them. On the other hand, it is very easy to forget about their existence and about the risk that someone somewhere may be watching. In RAVE, and in other media spaces, even seasoned users occasionally got confused about the properties of AV connections (Gaver 1992). Sometimes people even forgot about the presence of a video connection and about their visibility to others altogether. For example, if a colleague with whom one had an office-share connection switched off his camera or moved out of shot, it was easy to forget that one might still be on camera and visible to him. (See Dourish et al. 1996 for an extended description of the phenomenon of long-term AV office-share connections.)

In order to reduce such concerns, we can provide feedback about the presence of a camera, such as a monitor placed beside it. However, this only suggests that a video signal is available. It cannot reveal the signal's distribution, or when or to whom the image is being sent, or whether the image is being recorded or processed in any way. This may be tolerable in private offices with cameras and microphones, where the owner has control over when the devices are connected or even to whom they are allowed to be connected. In public areas, however, people may have no knowledge of the presence, capabilities, or purpose of such devices.

Concerns about privacy in such an environment can be attributed to the fact that the social space within which people operate is no longer identical to the physical space that is readily perceived (Harrison and Dourish 1996). The boundaries between private and public places become blurred; it may be impossible to judge who, if anyone, is virtually present and what they are capable of. This can be a major source of unease for people in that space, since the place they think they are in does not afford reliable cues as to how they should behave in terms of revealing information about themselves to others. In order to explain this unease, one must turn to considerations of the moment-to-moment control that people exercise over how they present themselves in public as social beings (Goffman 1963). In public places people actively control their privacy, behaving in a fashion that they think befits the setting.

People, especially newcomers and visitors to places with media spaces (which I take to imply awareness services) and other kinds of ubiquitous computing technology, can feel uneasy about their ability to monitor and control their self-presentation and consequently their privacy. This is because they have too little information about the nature of the technology, what it affords, and the character and intentions of the people using it. They have no reason to trust that a media-space system might not be capable of and even used for insidious purposes.

From a design perspective we can usefully think about these privacy concerns in terms of two phenomena which describe, in a cognitively significant sense, what is lost in social interactions mediated by technology: *disembodiment* from the context into which one projects and from which one obtains information (Heath and Luff 1991) and *dissociation* from one's actions. These phenomena interfere directly with the control of inflow and with the outflow of information, both of which are crucial to the management of privacy.

In the presence of others in face-to-face situations, you convey information in many ways. These include location, posture, facial expression, speech, voice level and intonation, and direction of gaze (Goffman 1959, 1963). Such cues influence the behavior of others. For example, they can determine whether others will try to initiate communication with you. Correspondingly, you read cues given by others about their state of mind, their availability for interaction, and so forth, and these influence your judgments about whether to attempt conversation, what to say, and how to act.

Disembodiment in CSCW and in ubiquitous computing environments means that these resources may be attenuated so that you may not be able to present yourself as effectively to others as you can in a face-to-face setting. For example, in an AV connection you may be only a face on a monitor (Gaver 1992), with your voice coming out of a loudspeaker whose volume you may not be aware of or able to control. You may appear only as a name in a list, as in our Virtual Café (see also McCarthy et al. 1991), or, in a system where people wear active badges that can be tracked by sensors around a building, you may appear only as a name associated with a room displayed on a screen (Harper et al. 1992). On the other hand, disembodiment also means that you may be unaware of when you are conveying information to others, because of a lack of feedback from the technology. For practical purposes, this means that it is difficult to tell if a person who is remotely connected to some location is really perceptible to others physically present in that location or if that person is able to perceive as fully as those physically present what is happening.

Dissociation means that a remote person may have no detectable presence at all (e.g., in a RAVE background connection). This occurs in CSCW applications when the results of actions are shared but the actions themselves are invisible or are not attributable to particular actor—in other words, when one cannot easily determine who is doing, or who did, what. Another example of this is ShrEdit, a synchronous, multi-user text editor that had no telepointers to indicate activities by remote co-authors and to show who was doing what (McGuffin and Olson 1992). Put simply, dissociation means that remotely connected people and their actions are impossible to detect or identify.

Breakdown of Social and Behavioral Norms and Practices

The effects of disembodiment and dissociation manifest themselves through a variety of violations of social norms and breakdowns in the use of CSCW and CMC systems. Disembodiment is reflected in a tendency for people to engage in prolonged, unintentional observation of others over AV links (Heath and Luff 1991). The intuitive principle "If I can see you, then you can see me" does not necessarily apply to computer-mediated situations, where one may be able to observe the activities of others without being observed. Further, people may intrude when they make AV connections, because they cannot discern how available others are (Louie et al. 1993).

Two major breakdowns related to dissociation are the inability to present oneself appropriately, not knowing if one is visible in some way to others, and the inability to respond effectively to a perceived action because one does not know who is responsible for it. A familiar example of this kind of breakdown arises with the telephone; it is not possible to identify nuisance callers before picking up the receiver.

Disembodiment and dissociation are, simply put, the most obvious manifestations of the shortcomings of CSCW, CMC, and ubiquitous computing technologies in terms of how they fail to support social interaction and how they foster unethical or irresponsible behavior. However, these shortcomings receive far less attention in the literature than insidious exploitation or subversion of technology. This is unfortunate, as they are highly important in mediated social interaction and communication and they are likely to be much more pervasive, particularly because they often relate to purely unintentional invasions of privacy. By addressing these shortcomings through careful design, we can reduce both inadvertent intrusions and the potential effects of subversive behaviors.

Design refinements and innovations can make the presence or the actions of remote collaborators available in such a way that they are accountable for their actions. Resources used in face-to-face situations can be exploited, simulated, or substituted. For example, media-space systems can embody the principle "If I see your face, you see mine," which is natural in face-to-face situations (Buxton and Moran 1990), or they can supply means to convey availability (Dourish and Bly 1992; Louie et al. 1993). Dissociation breakdowns in CSCW systems have been reduced by means of conveying identity, gestures, or even body posture (Minneman and Bly 1991; Tang and Minneman 1991).

A Design Framework

In general, disembodiment and dissociation in media spaces and other CSCW systems may be reduced through the provision of enriched feedback about information being captured and projected from users or their work to others. Users must also have practical mechanisms of control over that personal information. I now present a framework for systematically addressing these issues.

Addressing the Breakdowns

Much of the mutual awareness we normally take for granted may be reduced or lost in mediated interpersonal interactions. We may no longer know what information we are conveying, what it looks like, how permanent it is, to whom it is conveyed, or what the intentions of those using that information might be. In order to counteract breakdowns associated with this loss, this framework proposes that systems must be explicitly designed to afford feedback and control for at least the following four potential user and system behaviors:

capture   What kind of information is being picked up? Candidates include voices, actual speech, moving video or frame-grabbed images (close up or not), personal identity, work activity and its products, including data, messages and documents.

construction   What happens to information? Is it encrypted or processed at some point? Is it combined with other information (and, if so, how)? Is it stored? In what form? Privacy concerns in ubiquitous computing environments are exacerbated by the fact that potential records of our activity may be manipulated and later used out of their original context. This leads to numerous potential ethical problems (Mackay 1995).

accessibility   Is one's information public, available to particular groups, available only to certain individuals, or available only to oneself? What applications, processes, and so on utilize personal data?

purpose   To what uses is information put? How might it be used in the future? The intentions of those who wish to use data may not be made explicit. Inferring these intentions requires knowledge of the person, the context, or the patterns of access and construction.

I now consider each of these four classes of user and system behavior in relation to the following two questions: What is the appropriate feedback? What is the appropriate control? We thus have eight design questions, which form the basis for a design framework (table 1) with which we can analyze existing designs and explore new ones with respect to a range of privacy issues. This framework is a domain-specific example of the "QOC" approach to design rationale in which design issues, couched as questions, are explicitly represented together with proposed solutions and their assessments. (For more details see MacLean et al. 1991 and Bellotti 1993.)

The issues in the cells in table 1 are not necessarily independent of one another. For instance, in order to be fully informed about the purpose of

Table 1
A framework for design for privacy in CSCW, CMC, and ubiquitous computing environments. Each cell contains a description of the ideal state of affairs with respect to feedback or control of each of four types of behavior.

	Feedback about	Control over
Capture	When and what information about me gets into the system.	When and when not to give out what information. I can enforce my own preferences for system behaviors with respect to each type of information I convey.
Construction	What happens to information about me once it gets inside the system.	What happens to information about me. I can set automatic default behaviors and permissions.
Accessibility	Which people and what software (e.g., daemons or servers) have access to information about me and what information they see or use.	Who and what has access to what information about me. I can set automatic default behaviors and permissions.
Purposes	What people want information about me for. Since this is outside the system, it may only be possible to infer purpose from construction and access behaviors.	It is not feasible for me to have technical control over purposes. With appropriate feedback, however, I can exercise social control to restrict intrusion, unethical use, and illegal use.

information use, one must know something about each of the other behaviors. Likewise, in order to appreciate access, one must know about capture and construction. Understanding construction requires knowing something about capture. Hence, design of appropriate feedback must acknowledge these interdependencies. Control over each of them may, however, be relatively independently designed.

For those concerned about privacy, and about the potential for subversion in particular, control over (and thus feedback about) capture is clearly the most important issue. Given appropriate feedback about what is being captured, users can orient themselves appropriately to the technology for collaboration or communication purposes and exercise appropriate control over their behavior or what is captured in the knowledge of possible construction, access, and purposes of information use.

Evaluating Solutions

This framework emphasizes design to a set of criteria, which may be extended through experience and evaluation. Although questions about what feedback and control to provide set the design agenda, the criteria represent additional and sometimes competing concerns that help us to assess and distinguish potential design solutions (MacLean et al. 1991). A set of explicit criteria acts as a checklist helping to encourage systematic evaluation of solutions. These criteria have been identified from experiences with the design and use of a range of ubiquitous computing services. Particularly important in the current set are the following eleven criteria:

trustworthiness   Systems must be technically reliable and instill confidence in users. In order to satisfy this criterion, they must be understandable by their users. The consequences of actions must be confined to situations that can be apprehended in the context in which they take place and thus appropriately controlled.

appropriate timing   Feedback and control should be provided at a time when they are most likely to be required and effective.

perceptibility   Feedback and the means to exercise control should be noticeable.

unobtrusiveness   Feedback should not distract or annoy. It should also be selective and relevant, and it should not overload the recipient with information.

minimal intrusiveness   Feedback should not involve information that compromises the privacy of others.

fail safety   In cases where users omit to take explicit action to protect their privacy, the system should minimize information capture, construction, and access.

flexibility   What counts as private varies according to context and interpersonal relationships. Thus mechanisms of control over user and system behaviors should be tailorable to some extent by the individuals concerned.

low effort   Design solutions must be lightweight to use, requiring as few actions and as little effort on the part of the user as possible.

meaningfulness   Feedback and control must incorporate meaningful representations of information captured and meaningful actions to control it, not just raw data and unfamiliar actions. They should be sensitive to the context of data capture and also to the contexts in which information is presented and control exercised.

learnability Proposed designs should not require a complex mental model of how the system works. Design solutions should exploit or be sensitive to natural, existing psychological and social mechanisms and behaviors that are easily learned by users.

low cost Naturally, we wish to keep costs of design solutions down.

The first seven criteria are especially relevant to protection of privacy. The final four are more general design concerns. Some will inevitably have to be traded off against one another in practical design contexts.

These criteria are somewhat different in character from codes of ethics and privacy principles which are defined as professionally binding constraints on practice or as legal guidelines. A good example of such ethical codes and principles can be found in the Organization for Economic Cooperation and Development's Guidelines on the Protection and Privacy of Transborder Flows of Personal Data. (See Forester and Morrison 1990, pp. 144–147, for a discussion of how this code came about and some of its implications.) These guidelines, which can equally well be applied within nation states, are as follows:

collection and limitation Data must only be obtained by lawful means, with the data subjects' knowledge or consent.

data quality Data collectors must collect only data relevant to their purposes, and such data must be kept up to date, accurate, and complete.

purpose specification At the time of collection, the purposes to which the data will be applied must be disclosed to the data subject, and the data will not be used for purposes beyond this.

use limitation The data is not to be disclosed by the collector to outsiders without the consent of the data subject unless the law otherwise requires it.

security safeguards Data collectors must take reasonable precautions against loss, destruction, or unauthorized use, modification, or disclosure.

openness Data subjects should be able to determine the whereabouts, use, and purpose of personal data relating to them.

individual participation   Data subjects have the right to inspect any data concerning themselves and the right to challenge the accuracy of such data and have it rectified or erased by the collector.

accountability   The collector is accountable to the data subject in complying with the above principles.

These principles—intended as a comprehensive set of guidelines for designers, administrators, and regulators—can be applied to an enormous range of aspects of system design and of the practices of system users.

Now let us compare these principles with the criteria listed above. The criteria are to be used in conjunction with design efforts to address the particular questions presented in our design framework (table 1). While codes of ethics or principles are defined to govern the potentially intrusive use of systems that capture data about or offer access to people, the framework and the design criteria suggest that a system's design, rather than rules about the practices of its users, should be the vehicle for addressing most of these issues. For example, the collection and limitation principle is enforced automatically if the subject is given feedback about capture. (It is highly unlikely that unlawful data collection would occur if the framework were adhered to, because subjects could tell when this was happening.)

The eleven criteria themselves are not sufficient for privacy design without the framework; however, the framework alone could yield unwieldy and unworkable designs without applying the criteria to evaluate answers to its questions. In contrast, codes of ethics and principles like those of the OECD represent ideal states of affairs about which claims can be made and contested in a court of law or some professional tribunal, but they do not help designers to determine what system properties will achieve them. Furthermore, they do not tend to address themselves to the usability implications of mechanisms designed to protect privacy. The design framework presented here sets out to do just these things for designers through its systematic application.

Applying the Framework: Feedback and Control for Video Data from a Public Area

I now consider RAVE and the Apple Virtual Café in terms of the above framework in order to reveal aspects of their design that can be refined to improve users' privacy. For the sake of brevity, I focus on just one aspect of these two systems: the use of a video camera in a public place to send video information to people in remote offices.

In RAVE, people could access the camera in the Commons either with short (glance) or indefinite (background) video-only connections. The video data could be sent to a frame grabber that took digitized video snapshots. These could be used by various services, including Portholes (which sent frame-grabbed images around the building or across the Atlantic to the users at PARC).

At Apple, the first prototype of the Virtual Café displayed images taken from the camera in the Oh La La coffee bar on a web page which people in their offices could view by using a web browser. A touch-screen monitor, used by the staff members to select items which they wanted to advertise on the web page, displayed whatever names people typed into the form on the authentication page which they encountered before going to the Virtual Café page. This was a first attempt to let the staff members see who was looking at them, but this monitor on the staff's side of the counter could not be seen by the people waiting in line for coffee.

Providing feedback and control mechanisms over video signals or images taken from these public areas is challenging; however, it is important, since these areas are used by people who do not necessarily expect to be seen by others who are not visible in these places (i.e., who are dissociated from the act of viewing). In situations like this, designers must rely on users' exercising behavioral rather than technical control over information that they give out. The reason for this is that in a public place it is very difficult to offer the kinds of access control available to people in a private office, where a workstation or a personal computer is available as an interface to technical access-control mechanisms built into a media space.

The privacy framework prompts the following questions, for which I describe existing or potential design solutions. (Questions and relevant criteria appear in italics.)

What feedback is there about when and what information about me gets into the system?

Existing Solutions in RAVE

confidence monitor   A monitor was positioned next to the camera to inform passersby when they were within range, and what they looked like.

This solution fulfilled the design criteria of being *trustworthy*, *meaning-ful*, and *appropriately timed*.

mannequin   In order to alert people to the presence of the camera, a mannequin (affectionately named Toby) was positioned holding the cam-era. Toby drew people's attention because he looked like another person in the room. Originally the camera was in his head, but this concealed it and some visitors thought it was deliberately being hidden. Toby was then positioned with a camera on his shoulder. This feedback was less *obtru-sive* than the confidence monitor (which could be distracting), however it was less *meaningful* because it didn't tell you whether the camera was on or whether you were in its view.

Alternative Solution for RAVE

movement sensors   One solution that might have supplemented the use of confidence monitors would have been to use infrared devices to alert peo-ple, either with an audio or visual signal, when they moved into the field of view of a camera. These would provide *appropriately timed* feedback about onset of capture of video information; however, in order to be *mean-ingful* they would have to draw attention to the monitor by the camera.

Existing Solutions in the Virtual Café

warning sign   A sign displaying information about the presence of the camera tells customers in the café about what the camera is being used for and encourages them to try the system for themselves. This solution fulfills the design criteria of being *low-cost*, *unobtrusive*, and *meaningful* in some senses, but it is not *low-effort* for users to have to read. Further, the particular sign we put up was not highly *perceptible*. Signs are also not *meaningful*, in the sense that users cannot tell from them exactly what the image that is going to be snapped and sent out looks like. Additional audio feedback of a camera click was designed to alert people to an im-pending snapshot, but unfortunately this was not *perceptible* in the rather noisy café.

Alternative Solution for Virtual Café

public installation   A second version of the Virtual Café server has been built by Apple designers. This is a public installation which is placed prominently at the head of the line. Users can walk up to this installation and touch the screen to see the image the camera is currently displaying. This provides *meaningful* feedback about the nature of the snapshots taken by the camera by displaying the most recent one. However, this information is not *appropriately timed*, since it is not available as the camera takes the picture. Because the screen must go into a screen-saver mode when nobody is interacting with it (to avoid burn-in of the rather static display), the image is not shown until a user touches the screen.

Proposed Solutions for Virtual Café

design refinements to signs and audio feedback   We could easily and cheaply improve on the design of our warning sign and our audio feedback to make them more *perceptible*.

What feedback is there about what happens to information about me inside the system?

No Existing Solution for RAVE

The confidence monitor did not inform visitors and newcomers to the lab that the video signal was sent to a switch that could potentially direct the signal to any number of nodes, recording devices, or ubiquitous computing services in the lab. Unfortunately, Toby could not tell you whether recording or some other construction of video data was taking place. In fact, EuroPARC's policy was that recording never took place without warnings to lab members.

Proposed Solutions for RAVE

LED display   A simple solution would have been to place an LED status display near the camera. Additional information would have had to be displayed to make the readings on the display *meaningful*. This would have been a *low-cost* proposal to give appropriately timed feedback about when and which services were actively collecting information. Unfortunately, it might not have been sufficiently *perceptible*.

audio and video feedback   Audio feedback could have been provided to indicate connections and image frame-grabbing, but it might have been *obtrusive* and annoying to provide repeated audio feedback to a public area at a sufficient volume to be heard all round the room. An alternative would have been to superimpose the flashing word "Recording" on the screen of the confidence monitor. This would have had to be designed to be *perceptible* yet *unobtrusive*.

Existing Solution for Virtual Café

warning sign   A sign in the Oh La La coffee bar explains the purpose of the camera and what happens to the images taken by the camera. Like the above-mentioned solutions, this is not *low-effort* or *perceptible*.

Proposed Solutions for Virtual Café

information display   As with the media space, the Oh La La installation itself could display *meaningful* information about what was happening to the video information. (No recording or processing of any kind takes place in Oh La La, so it would not be necessary to have additional feed-back about processing of information beyond feedback about pictures being taken.) Unfortunately, because the system is usually in the screen-saving mode, feedback would not be *appropriately timed*, nor would it be *low-effort* for users to obtain.

What feedback is given about who has access to information about me and what information they see?

Existing Solutions for RAVE

textual information   In order to warn people that they might be watched, Toby wore a sweatshirt with the message "I may be a dummy but I'm watching you!" printed on it. Since Toby was holding a camera, the meaning of this was fairly clear in a general sense, but it was not *meaningful* enough in terms of who was really doing the watching and when.

Proposed Solutions for RAVE

viewer display   One option would have been to display a list of names or pictures on the wall to indicate who was watching a public area. For *appropriately timed*, updated information, it would have been possible to adapt Portholes for this purpose. In order to be *perceptible* to passersby, the display would have had to be larger than a normal monitor. We could have projected images of who was currently connected to the camera onto the wall. However, this would not have been *low-cost*, and might have seemed *intrusive* to those who connected to this public area.

audio feedback   In private offices, audio feedback alerted occupants to onset and termination of short-term glance connections. Such feedback about video connections to the Commons would not have been *appropriately timed*, since they were normally of the long-term background variety (which outlasted most visits to the room).

Existing Solutions for Virtual Café

passport ID   The first version of the Virtual Café only displayed a list of the names users chose to type into the authentication page on the web (which could be anything—people would sometimes type things like "The Clown" and "A spy"). The second version of the system does not give people the option to visit in different disguises whenever they feel like it. Instead, users are invited to take a picture ID using the camera in the Oh La La coffee bar. When one first downloads the client software for the system, one types in one's name and password, and a photo ID is automatically downloaded. Each time one starts the client after this, one's ID is displayed on all clients of the server, including the public one in Oh La La.

Although a user could type in a false name and not stand in front of the camera when it took his or her photo, this seems highly unlikely. We have yet to establish exactly what patterns of use emerge in this respect. The main point here is that now people can be recognized from their pictures as well as from their names, and these pictures can be seen by anyone who walks up to the installation. This provides *meaningful* feedback about who if anyone is currently looking at the Oh La La picture, though it is clearly not completely *trustworthy*.

Proposed Solutions for Virtual Café

There is no simple way of overcoming the possibility of dissociation between users and their actions (i.e., failure of authentication). It would be difficult to guarantee that we would know who was looking (though we can be more confident about how many are looking). If we became concerned about people trying to defeat the passport ID system, we could insist that people type in their employee numbers along with their names, and we could cross-check these details, refusing access to those who typed in false names. Under the circumstances this seems rather farfetched, though this might be a more *trustworthy* solution for other less informal contexts where such a system might be installed.

What feedback is provided about the purposes of using information about me?

RAVE and Virtual Café: No Existing Solution

There is no technical feedback in either of these systems about the intentions of those who access the video signals or images. However, lab members at EuroPARC were reasonably confident that their privacy was not threatened by others' uses of information, probably because, in a public area, they already behaved in what they considered to be a publicly presentable and acceptable fashion. It seems reasonable to assume that the same might be said of those at Apple, although a preliminary study suggested that a few people who frequent Oh La La dislike being on camera regardless of the intentions of those who might be seeing them.

Proposed Solution

Unfortunately, though we might wish to, we cannot provide technological feedback about people's intentions. We can only deduce intentions from feedback about capture, access, and utilization of information, together with knowledge of or interactions with the culture, the context, and the individuals concerned.

What control is there over when and when not to give out what information?

Existing Solutions for RAVE

moving off camera   People could move on or off camera; there was a clear video-free zone around a corner in the Commons for those who wanted to relax in private. This kind of solution is very *easy to learn*.

covering the camera   If certain activities, such as the potentially embarrassing weekly step-aerobics workout, were taking place in the Commons, the camera's lens was covered. People felt they could do this because, in this context, they felt some ownership of the system and thus felt that they could tamper with the technology with impunity when they saw fit. This was therefore a highly *flexible* and *trustworthy* solution.

behaving appropriately   If people had to walk in front of the camera, they could orient themselves appropriately to it, given the *meaningful* feedback they obtained from the confidence monitor. (Those familiar with the system did not go right up to the camera and stare into it, as some newcomers did, for fear of appearing foolish to remote viewers.) Behaving as though one is on public display is certainly *trustworthy, low-effort, learnable,* and *low-cost.* However, if there are no other means of control, it means that there is no *flexibility* for changing the technical setup and engaging in private behaviors, and the presence of a camera may feel highly *intrusive* to those concerned.

Existing Solutions for Virtual Café

behaving appropriately   The only realistic resource for the customers at Oh La La is for them to behave as if they might be being photographed. As long as they are aware of the camera, this is a *trustworthy, low-effort, learnable, low-cost* solution. However, since the camera points at the line for coffee, they cannot easily avoid it, nor is there the same sense of ownership of the technology that exists when people feel that they can cover the lens. *Flexibility* is lost, and it is hardly surprising that a preliminary study revealed that some feel the presence of a camera in this context to be *intrusive.*

What control is there over what happens to information, who can access it and for what purposes it is used?

Neither RAVE nor the Apple Virtual Café currently offers technical control over the capture, construction, and access of video signals, because each of them is a public rather than a private information source. Technical control over the purposes for which information can be used, however, is impractical with any information source, public or private. At EuroPARC and at Apple, social control is effectively exercised over such systems. (There were many debates and meetings at both sites on the subject of what was and was not acceptable.) The culture of acceptable use that evolved with RAVE (a culture one would expect to see evolve with the Virtual Café also) dictated that certain practices were frowned upon.

At EuroPARC people were always expected to warn others that constructions and access other the normal background connection, glance, and Portholes services were occurring. For example, experiments involving a video diary (Eldridge et al. 1992), in which subjects had their pictures taken wherever they went, were announced to all lab members, and the subjects wore labels warning others of potential recording of camera images in areas they entered. Anyone who objected to this could either avoid the label wearers or ask the researchers concerned to make sure that they were not recorded.

Use of the framework

In the examples I have given, the design framework serves three important purposes:

It helps to clarify the existing state of affairs with respect to privacy concerns and with respect to the social norms and practices currently in place.

By clarifying the system and user behaviors that give rise to these concerns, it helps to inspire a range of possible design solutions.

It is used to assess the solutions in terms of how they might reduce the existing concerns as well as how they might give rise to new ones.

With regard to the past purpose, it is important to point out that, while the framework and the design criteria help us evaluate proposed solutions prospectively, it is only by implementing these solutions that we can truly judge their usefulness. A good example of this is the RAVE software's infrastructure, Godard, whose privacy feature allowed users to select who was permitted to connect to them. Consistent with the criterion of being fail safe, the default was that a newcomer to the lab was automatically denied access until each person in the lab explicitly allowed him or her access. Unfortunately, because lab members generally forgot to change their settings, newcomers found that they were unable to connect to anyone. As a result, they often gave up trying and thus never got into the habit of using RAVE. There are simple potential solutions to this, such as automatically notifying people to whom an unsuccessful attempt to connect has been made. However, many such difficulties are not foreseen, so evaluation through use must always be part of the design cycle.

Discussion and Conclusions

Privacy can be defined as a capability to determine what one wants to reveal and how accessible one wants to be. This notion is tied to the design challenges inherent in media spaces and similar systems which blur the boundaries between private and public space and in which disembodiment and dissociation make it difficult to know how to manage one's privacy and how to behave. Such systems create virtually connected places with the possibility of remote but invisible mediated presences. Having offered a design framework for those who wish to tackle these kinds of challenges explicitly, I will conclude by reflecting on some wider issues concerning designing for privacy and the possible use of this framework.

How Far Should We Go?

The framework for design for privacy, together with its associated design criteria, is intended to help designers of CSCW systems, CMC systems, and ubiquitous computing systems to ensure that their users do not suffer from the above-mentioned disembodiment and dissociation. Disassociation can be overcome to a great extent by providing resources that enable people to determine who is looking at them or otherwise accessing and constructing information about them in some way. Disembodiment is clearly a more imposing challenge, as it is technologically difficult to embody people in remote places. (We do not yet have the affordable immersive virtual reality and holographic technologies that we would need to even approach remote embodiment.)

The question we must ask, therefore, is how far do we need to go in order to protect privacy in potentially intrusive CSCW systems, CMC systems, and ubiquitous computing systems. From a design perspective, for all practical purposes, privacy is not simply keeping some particular information from others, nor is it preserving some state of affairs. To support privacy in the design of potentially intrusive systems (such as media spaces) is to provide users with the capability to understand what information they are revealing and how, and to be able to control this disclosure. People can understand what they are revealing only if they are provided with appropriate feedback through well-designed systems. Control may be exercised through technical means, such as blocking certain kinds of system-mediated actions, or through the kinds of behaviors in which people engage in public places.

In private offices, individual users of systems like media spaces can decide whether to allow certain kinds of connections or whether to have a camera pointed at them. In public places, as in the examples given above, it is more difficult to offer technical control over potentially intrusive systems, such as those that deploy cameras or microphones, since people may have different feelings about what is and what is not private and may be unfamiliar with how to control the technology anyway. Consequently, the simplest means of dealing with privacy issues in public settings is to provide rich feedback, and indeed the application of the framework highlights this point quite clearly. The designers of the RAVE and Virtual Café systems were not able to come up with means of offering technical control over information capture for people in public places. On the other hand, they had many ideas about how to provide suitable feedback.

Thus, perhaps the best we can do in practice is give *feedback about actual or potential information capture in a perceptible but unobtrusive fashion*. This would be enough to alert people to the need to exercise technical control, or, in public places, to behave appropriately. However, it may be useful to provide feedback about when information is accessed and how it is constructed and to represent the identities of individuals accessing information in order to ensure that *people are always accountable for their actions*. This latter course of action is certainly closer to the ideal, but it would also be more difficult and expensive.

In most circumstances, therefore, we might expect that the absolute *least* designers should do is *ensure that devices that capture people's activities are designed to convey that this is what they do*, whether or not they are connected to recorders or to processing systems or directly to other people. This is not necessarily expensive, as some of the solutions outlined above should suggest, and it would be a valuable safeguard against individuals' assuming that they are in a private setting when they are not. Insisting on this kind of system property in certain contexts could easily be seen as a basis for policy or law in relation to the rapidly growing use of CMC and surveillance systems.

How Far Can We Generalize?

In this chapter I have focused on technical design features, rather than on social or policy means of protecting privacy in ubiquitous computing environments. I have argued that this is an important design focus, and have offered a framework as guidance. That being said, it is interesting to reflect on the ways in which cultural norms and practices affect and sometimes ameliorate privacy concerns. This is particularly well illustrated by experiences with extended use of the RAVE media space.

In spite of shortcomings which the framework highlights, RAVE and the Apple Virtual Café were, in general, highly acceptable to most members of the respective communities of use. In part, this was because feedback and control of information capture were well attended to overall in their design, enabling people to orient themselves appropriately to the technology. Further, people knew roughly how the systems worked, and trusted the benign cultures which governed their use. However, one can imagine that, in other contexts, a media space could seem very sinister indeed. It is thus quite understandable that many visitors to EuroPARC and to Apple experienced some trepidation.

RAVE and the Apple Virtual Café are research tools, as are most ubiquitous computing systems at present. Neither of these is intended to be a model for a commercial, standard system. Rather, they are instances of prototypes' evolving with a culture of use and thus being acceptable only to members of their respective cultures. The concerns highlighted by the privacy-design framework could, however, point to refinements that might be essential in other contexts.

Privacy in the Balance: Some Further Issues
Another issue which the framework helps to elucidate is the delicate balance that exists between awareness and privacy. Providing too much awareness of other people's activities and availability may be seen as intrusive. However, too little awareness may result in inadvertent invasions of privacy, such as when people cannot tell how receptive an individual is to being disturbed. The ideal is to provide awareness without crossing the line into intrusiveness.

A related issue is that data that could be feedback to one person may include personal information about another person which that person might not want to make available. For example, the proposal to display large Portholes images of people connected to the Commons camera in RAVE could make them feel as if they are on display. This suggests that designs which address some privacy problems may give rise to others. Our framework should therefore be applied not only to the systems in place but also to the privacy solutions themselves. In this way it can be used to highlight design tradeoffs in adopting alternative solutions.

It is often the case that design decisions involve tradeoffs and compromises. Designing to protect privacy is extreme in this respect, since one person's awareness can imply that another's privacy is being compromised. There is no guarantee that features put in place to protect privacy will confer the benefits that the designer intends and will not cause new difficulties. The framework described above aims to provide a systematic basis for tackling a range of user-interface design challenges. However, in view of the tradeoffs and the unforeseen consequences of design features, there is still no substitute for *in situ* evaluation and iteration.

A final point I wish to make is that particular concerns about privacy tend to wax and wane over time. Privacy guidelines tend to show their age very rapidly as the capabilities and applications of computing technology evolve in unpredictable and radical ways. For example, guidelines often include a variation of the OECD's purpose-specification principle which is based on concerns about organizations' gathering and selling or otherwise distributing financial, medical, criminal, or other potentially personal records from databases to be used for indeterminate purposes. However, ubiquitous computing systems as envisioned by Weiser (1991) and as implemented in applications of EuroPARC's RAVE system (e.g., as a video diary; see Eldridge et al. 1992) almost by definition cannot comply with such a principle. The point and the advantage of the technology is that information is gathered as a resource, and *the purpose of the data cannot be specified until it is used*, if at all, at some unpredictable time in the future.

As guidelines begin to show their age, attitudes toward privacy have changed. New technologies with attractive benefits, such as cellular telephones, tend to be accepted as a worthwhile tradeoff by their purchasers. It is worth having one's location recorded in a way not previously possible, and it is worth the increased intrusion from unwanted calls, to get the benefit of being able to communicate by phone wherever one is. With time, people may expect to be locatable at all times, or they may develop increased sensitivity to unwanted intrusions. Privacy will continue to be balanced against other concerns of modern life, and it will continue to be a matter of feedback and control; however, it may not be possible to predict what people will care to have feedback about and what they will choose to exercise control over twenty years from now.

Acknowledgements

I am greatly indebted to Abi Sellen, with whom many of the concerns and ideas in this chapter were first conceived and who assisted in the preparation of a paper on which this chapter is based (Bellotti and Sellen 1993). Also I thank Paul Dourish, Mike Molloy, Bill Walker, Charlie Hill, and Marion Buchenau, who are responsible for collaborating on and implementing many of the design ideas. I respectfully thank Steve Harrison for his insights into the significance of place. I also thank Bob Anderson, Sara Bly, Graham Button, Matthew Chalmers, Paul Dourish, Bill Gaver, Steve Harrison, Mik Lamming, Paul Luff, Wendy Mackay, Allan MacLean, Scott Minneman, William Newman, Peter Robinson, and Dan Russell for support, interesting discussions, helpful comments, and indulgence.

Notes

1. This is based on Rule's (1980) distinction between violations of aesthetic privacy (which expose things that victims may feel are inappropriate to reveal to others) and violations of strategic privacy (which may compromise the victim in the pursuit of his or her interests).
2. EuroPARC has since been renamed "Rank Xerox Research Center, Cambridge Laboratory." It is located in Ravenscroft House.

References

Bellotti, V. 1993. Integrating theoreticians' and practitioners' perspectives with design rationale. In Proceedings of INTERCHI93 (Conference on Human Factors in Computing Systems), Amsterdam.

Bellotti, V. 1996. What you don't know can hurt you: Privacy in collaborative computing. In *People and Computers XII*. Springer-Verlag.

Bellotti, V., and P. Dourish. 1997. Rant and RAVE: Experimental and experiential accounts of a media space. In *Video-Mediated Communication*, ed. K. Finn et al. Erlbaum.

Bellotti, V., and A. Sellen. 1993. Design for privacy in ubiquitous computing environments. In Proceedings of the European Conference on Computer-Supported Cooperative Work, Milan.

Bowyer, K. 1996. *Ethics and Computing: Living Responsibly in a Computerized World*. IEEE Computer Society Press.

Buxton, W., and T. Moran. 1990. EuroPARC's integrated interactive intermedia facility (IIIF): Early experiences. Presented at IFIP Conference on Multi-User Interfaces and Applications, Herakleion, Crete.

Chaum, D. 1992. Achieving electronic privacy. *Scientific American* 267, August: 96–101.

Clarke, R. 1988. Information technology and dataveillance. *Communications of the ACM* 31, no. 5: 498–512.

Cool, C., R. Fish, R. Kraut, and C. Lowery. 1992. Iterative design of video communication systems. In Proceedings of the ACM Conference on Computer-Supported Cooperative Work, Toronto.

Denning, D., and D. Branstad. 1996. A taxonomy for key escrow encryption systems. *Communications of the ACM* 39, no. 3: 34–40.

Dourish, P. 1991. Godard: A Flexible Architecture for A/V Services in a Media Space. Technical Report EPC-91-134, Rank Xerox EuroPARC, Cambridge.

Dourish, P. 1993. Culture and control in a media space. In Proceedings of the European Conference on Computer-Supported Cooperative Work, Milan.

Dourish, P., and S. Bly. 1992. Portholes: Supporting awareness in distributed work groups. In Proceedings of the ACM Conference on Human Factors in Computing Systems, Monterey, California.

Dourish, P., A. Adler, V. Bellotti, and A. Henderson. 1996. Your place or mine? Learning from long-term use of video communication. *Computer-Supported Cooperative Work* 5, no. 1: 33–62.

Dunlop, C., and R. Kling, R. 1991. *Computerization and Controversy: Value Conflicts and Social Choices*. Academic Press.

Eldridge, M., M. Lamming, and M. Flynn. 1992. Does a video diary help recall? In People and Computers VII, Proceedings of the HCI92 Conference, York.

Fish, R., R. Kraut, and R. Root. 1992. Evaluating video as a technology for informal communication. In Proceedings of the ACM Conference on Human Factors in Computing Systems, Monterey, California.

Forester, T., and P. Morrison, P. 1990. *Computer Ethics*. MIT Press.

Gaver, W. 1991. Sound support for collaboration. In Proceedings of the European Conference on Computer-Supported Cooperative Work, Amsterdam.

Gaver, W. 1992. The affordance of media spaces for collaboration. In Proceedings of the ACM Conference on Computer-Supported Cooperative Work, Toronto.

Gaver, W., T. Moran, A. MacLean, L. Lövstrand, P. Dourish, K. Carter, and B. Buxton. 1992. Realising a video environment: EuroPARC's RAVE system. In Proceedings of the ACM Conference on Human Factors in Computing Systems, Monterey, California.

Goffman, E. 1959. *The Presentation of Self in Everyday Life*. Anchor .

Goffman, E. 1963. *Behavior in Public Places*. Free Press.

Harper, R., M. Lamming, and W. Newman. 1992. Locating systems at work: implications for the development of active badge applications. *Interacting with Computers* 4: 343–363.

Harrison, S., and P. Dourish. 1996. Re-placing space: The roles of place and space in collaborative systems. In Proceedings of the ACM Conference on Computer-Supported Cooperative Work, Boston.

Harrison, S., S. Bly, S. Minneman, and S. Irwin. 1997. The media space. In *Video-Mediated Communication*, ed. K. Finn et al. Erlbaum.

Heath, C., and P. Luff. 1991. Disembodied conduct: Communication through video in a multi-media office environment. In Proceedings of the ACM Conference on Human Factors in Computing Systems, New Orleans.

Heath, C., and P. Luff. 1992. Collaboration and control: Crisis management and multimedia technology in London underground line control rooms. *Computer-Supported Cooperative Work* 1, no. 1–2: 69–94.

Hindus, D., and C. Schmandt. 1992. Ubiquitous audio: Capturing spontaneous collaboration. In Proceedings of the ACM Conference on Computer-Supported Cooperative Work, Toronto.

Lamming, M., and W. Newman. 1991. Activity-Based Information Retrieval Technology in Support of Personal Memory. Technical Report, EPC-91-103.1, EuroPARC.

Lampson, B., M. Paul, and H. Siegert. 1981. *Distributed Systems: Architecture and Implementation.* Springer-Verlag.

Louie, G., M. Mantei, and A. Sellen. 1993. Making Contact in a Multi-Media Environment. Unpublished manuscript.

Mackay, W. 1995. Ethics, lies and videotape. In Proceedings of the ACM Conference on Human Factors in Computing Systems, Boston.

MacLean, A., R. Young, V. Bellotti, and T. Moran. 1991. Questions, options, and criteria: Elements of a design rationale for user interfaces. *Human-Computer Interaction* 6, no. 3–4: 201–250.

Mantei, M., R. Becker, A. Sellen, W. Buxton, T. Milligan, and B. Wellman. 1991. Experiences in the use of a media space. In Proceedings of the ACM Conference on Human Factors in Computing Systems, New Orleans.

McCarthy, J., V. Miles, and A. Monk. 1991. An experimental study of common ground in text-based communication. In Proceedings of the ACM Conference on Human Factors in Computing Systems, New Orleans.

McClurg, A. 1995. Bringing privacy law out of the closet: A tort theory of liability for intrusions in public places. *North Carolina Law Review* 73, no. 3: 989–1088.

McGuffin, L., and G. Olson. 1992. ShrEdit: A Shared Electronic Workspace. Technical Report, Cognitive Science and Machine Intelligence Laboratory, University of Michigan.

Milberg, S., S. Burke, H. Smith, and E. Kallman. 1995. Values, personal information, privacy and regulatory approaches. *Communications of the ACM* 38, no. 12: 65–74.

Minneman, S., and S. Bly. 1991. Managing à trois: A study of a multi-user drawing tool in distributed design work. In Proceedings of the ACM Conference on Human Factors in Computing Systems, New Orleans.

Minneman, S., S. Harrison, W. Jansson, G. Kurtenbach, T. Moran, I. Smith, and W. van Melle. 1995. A confederation of tools for capturing and accessing collaborative activity. In Proceedings of Multimedia95 (ACM, New York).

Mullender, S. 1989. Protection. In *Distributed Systems*, ed. S. Mullender. Addison-Wesley.

Parker, D., S. Swope, and B. Baker. 1990. Ethical Conflicts in Information and Computer Science, Technology, and Business. QED.

Pedersen, E., K. McCall, T. Moran, and F. Halasz. 1993. Tivoli: An electronic whiteboard for informal workgroup meetings. In Proceedings of INTERCHI93 (Conference on Human Factors in Computing Systems, Amsterdam).

Privacy Protection Study Commission 1991. Excerpts from *Personal Privacy in an Information Society*. In *Computerization and Controversy: Value Conflicts and Social Choices*, ed. C. Dunlop and R. Kling. Academic Press.

Reiman, J. 1995. Driving to the panopticon: A philosophical exploration of the risks to privacy posed by the highway technology of the future. *Santa Clara Computer and High Technology Law Journal* 11, no. 1: 27–44.

Root, R. 1988. Design of a multi-media vehicle for social browsing. In Proceedings of the ACM Conference on Computer Supported Cooperative Work, Portland, Oregon.

Rule, J. B., et al., eds. 1980. *The Politics of Privacy: Planning for Personal Data Systems as Powerful Technologies*. Elsevier.

Samarajiva, R. Interactivity as though privacy mattered. In present volume.

Smith, R., T. O'Shea, C. O'Malley, E. Scanlon, and J. Taylor. 1989. Preliminary experiments with a distributed, multi-media, problem solving environment. In Proceedings of the European Conference on Computer-Supported Cooperative Work, London.

Stone, E., D. Gardner, H. Gueutal, and S. McClure. 1983. A field experiment comparing information-privacy values, beliefs and attitudes across several types of organizations. *Journal of Applied Psychology* 68, no. 3: 459–468.

Stults, R. 1988. The Experimental Use of Video to Support Design Activities. Report SSL-89-19, Xerox PARC, Palo Alto.

Tang, J., and S. Minneman. 1991. VideoWhiteboard: Video shadows to support remote collaboration. In Proceedings of the ACM Conference on Human Factors in Computing Systems, New Orleans.

Tang, J., E. Isaacs, and M. Rua. 1994. Supporting distributed groups with a montage of lightweight interactions. In Proceedings of the ACM Conference on Computer Supported Cooperative Work, Chapel Hill, North Carolina.

Want, R., A. Hopper, V. Falcco, and J. Gibbons. 1992. The active badge location system. *ACM Transactions on Office Information Systems* 10, no. 1: 91–102.

Weisband, S., and B. Reinig. 1995. Managing user perceptions of email privacy. *Communications of the ACM* 38, no. 12: 40–47.

Weiser, M. 1991. The computer for the 21st century. *Scientific American* 265, September: 94–104.

3

Convergence Revisited: Toward a Global Policy for the Protection of Personal Data?

Colin J. Bennett

In *Regulating Privacy*, published in 1992 and based on an analysis of data protection and privacy in the 1970s and the 1980s, I concluded that strong pressures for "policy convergence" had forced different states to legislate a broadly similar set of statutory principles to grant their citizens a greater control over personal information. The general pattern of policy development seemed to support the hypothesis that globalizing forces for economic and technological development were progressively overwhelming the distinctive institutional capacities, historical legacies, and cultural traditions of advanced industrial states.[1] Countries that wanted an "information privacy" or "data protection" policy had a limited repertoire of solutions from which to choose. They each selected a version of the familiar "fair information principles."

States did not reach this convergence of policy because of an inexorable technological determinism, however. Certainly the rapidly evolving and pervasive information technologies shaped the structure of the debates in each country. But they did not predetermine policy outcomes. Rather, the pattern was one in which a group of experts in the 1960s and early 1970s forged *ad hoc* contacts in order to discuss the intrusive effects of new information technologies. Over time a cross-national policy community coalesced around a shared concern for the privacy problem and a desire to emulate the policies of other jurisdictions.

As communications intensified throughout the 1970s, there arose a broad recognition of a commonality of concern, and pressure was placed upon international organizations to codify this consensus in the nonbinding guidelines of the Organization for Economic Cooperation and Development[2] and in the full convention of the Council of Europe.[3]

These agreements marked an early recognition of the need for the harmonization of data-protection policy in an era of increasing international data flows. During the 1980s, therefore, data protection was more properly regarded as an international issue on an international agenda. Policy makers that have legislated data protection more recently have needed to pay more and more attention to the agreements reached at the international level, with the result that a more *penetrative* process is observed in these countries in the 1990s.

The development of privacy and data-protection policy was thus examined according to this theoretical framework for policy convergence (technological determinism, emulation, elite networking, harmonization, and penetration). I concluded that these different forces had variable impacts over time:

For the pioneers, the United States and Sweden, the convergence resulted from independent and indigenous analyses that traveled along the same learning curve and arrived at the same conclusion. For West Germany, and other countries such as Canada, France, Norway, Denmark and Austria that legislated in the late 1970s, the convergence followed from the mutual process of lesson drawing within an international policy community. For Britain, and other laggards such as the Netherlands, Japan, and Australia, the convergence has resulted from the pressure to conform to international standards for mainly commercial reasons.[4]

These forces generally overwhelmed differences in formal constitutional frameworks, legal traditions, partisan ideologies and political cultures.

At the end of the 1980s, three major divergences were observed in the world's data-protection policies. The first concerned scope. Most of the European countries applied the same statutory principles to both the public and the private sector. The United States, Canada, Australia, and Japan, however, rejected an "omnibus" approach, preferring to regulate only the public sector's practices and to leave the private sector governed by a few sectoral laws and voluntary codes of practice.[5] The second difference reflected a disagreement about whether the laws should apply only to computerized (automated) personal data or also to manual record-keeping systems. A majority of countries chose to make no distinction; exceptions include Sweden, the United Kingdom, and Austria.

The third and principal difference concerned the choice of policy instruments to enforce, oversee, and administer the implementation of the legislation. Data protection may be administered through a range of tools that give variable responsibilities to individual citizens and a variety of advisory and regulatory powers to data-protection agencies. These powers range from the stricter licensing and registration regimes in force in Sweden and Britain to the more advisory and less regulatory systems headed by privacy or data-protection "commissioners" in Germany, Canada, and Australia.[6] To some extent the choice of these instruments was informed by historical legacies that shaped various countries' notions of the "natural" way to enforce administrative law. Each country had a distinct set of "policy equipment" or a "policy toolkit" from which these different instruments were chosen.[7]

One of the assumptions behind this analysis was that "information privacy" or "data protection" were considered as discrete legal or technological issues. At the time, I was critical of much of the literature for failing to locate and understand the policy problem in the context of wider sociological and political trends. I contended that there was enormous scope for political scientists to direct their theoretical and conceptual tools to this subject.[8] As an example of regulatory policy, data protection raises fascinating questions about the capacities of different political systems to manage technological change in order to address one of the fundamental claims of modern citizenship—the "right to be let alone."

Regulating Privacy was written at a time when only seventeen advanced industrial OECD countries had produced data-protection laws based on the model of "fair information principles." My analysis in that book was based on a more intensive investigation into the politics behind four of these laws: the 1973 Swedish Data Act, the 1974 American Privacy Act, the 1977 West German Data Protection Act, and the 1984 British Data Protection Act. Since then, nearly all European countries have joined the "data-protection club," including some former Communist states (e.g. Hungary). Moreover, the governments of Canada and Australia have pledged to extend their privacy-protection policies in the public sector to the private. Data protection around the world is now extensive. In the advanced industrialized world, countries with laws clearly outnumber those without.

The purpose of this chapter is to extend the analysis presented in *Regulating Privacy* to the 1990s and beyond. A range of more recent technological, economic, and political developments raise interesting questions about whether the patterns of convergence and divergence identified in the late 1980s are still observed. To what extent has there been a continuing convergence of public policy? If so, which of the forces for convergence provides the best explanation? What further developments have created pressures for national divergence? In order to address the potential for a truly global data-protection policy, I would argue, we must continue to grapple with the different "forces for convergence" and to understand their relative influence on data-protection policy in the 1990s and beyond.

Forces for Policy Convergence in the 1990s

Just as divergence is not a synonym for difference, convergence is not a synonym for similarity. Rather, it suggests a dynamic process by which countries begin at different starting points and over time converge toward similar solutions. The critical dimension is, therefore, a temporal rather than a spatial one.[9] The analysis must also be specific about what aspects of public policy are supposed to be converging. A distinction among policy content (the principles and goals of the public policy), policy instruments (the institutional means by which those principles are implemented), and policy outcomes (effects on individual and organizational behavior) will prove useful for the purposes of this analysis. To what extent and in what respects, therefore, has the world's data-protection policy been converging?

National and Global Information Highways: The Technological Imperative

Regulating Privacy was written before the initiation of attempts in various advanced industrial states to fashion appropriate technological, economic, and regulatory contexts for the development of interactive "information highways" or "information infrastructures." The technologies are now more interactive and less centralized. Nevertheless, the same dynamics are at work as existed in the 1960s with respect to the mainframe "databanks" to which the initial data-protection laws were directed. All information and communications technologies have some intrinsic properties that have forced policymakers in different countries to treat their dangers in a similar manner.

The first characteristic is obviously pervasiveness. New information technologies have always diffused around the advanced industrial world through global capitalism with few if any restrictions. Unlike the resources upon which the industrial economies were based (capital, labor, natural resources), the principal commodity of the information age knows few national attachments. The associated problems are therefore likely to be commonly recognized and similarly treated.

The more discrete mainframe "databanks" of the 1960s and the 1970s served to frame a similar set of problems associated with the processing of similar forms of personal information, but in discrete and largely unconnected locations. The development of global information networks has changed and intensified the character of the privacy-protection problem. No longer can personal information be said to be located anywhere. It is susceptible to instantaneous transmission anywhere in the world. Organizational "databanks" are now parts of national "information highways," which in turn are gateways to a ubiquitous global information infrastructure.[10]

In the 1960s and the 1970s, the privacy challenge could be nationally conceived and addressed. It so happened that elites in the pioneering countries interacted constantly with one another and resolved to address the problem in roughly similar ways. In the 1980s and the 1990s, because of the ubiquity, ease of use, speed, and decentralization of the technology, privacy is unequivocally an international issue on an international agenda. One's privacy is now less threatened by the omniscient gaze of a centralized "Big Brother" than by the unknown and unseen collection, matching, and profiling of transactional data, a trail of which is left by every one of us as we purchase goods, apply for services, make entertainment choices, and so on. The "new surveillance" is decentralized, routine, increasingly global, and even more likely to have a convergent influence on public policy.

A second characteristic of these technologies is their rapid development, or at least the rapid development in distributed access to distributed databases. This has the following policy consequences: Policymakers have always faced the likelihood that their regulatory efforts will be obsolete by the time they are implemented. The making of privacy policy takes place in anticipation of largely unknown technological developments and dangers, producing a strong incentive to formulate laws with sufficient latitude to embrace future eventualities. The generality of the "fair information principles" is thus a "reflection of the need for a set of basic rules of sufficient breadth and adaptability."[11]

It might be argued that changes in computing technology are superficial and that the "deep grammar" of computing practice displays some striking continuities.[12] Even though computing is based on some essential practices and norms that display a consistency over time, legislators in national and international arenas have always had to operate on a ground that is moving beneath their feet. Perhaps the dynamics today are more intense than they were 20 years ago. The rapidity of change will continue to call for policies that are stated in terms of some very general principles that can be applied to the technology practices within our public and private organizations. Indeed, as we shall see below with the European Union's Data Protection Directive, some more specific distinctions that were made in previous legislation have now given way to less precise formulations that are arguably more sensitive to the new environment.[13]

Information technology has been described as mysterious, complex, or enigmatic. Laws and regulations have typically been made and enforced by men and women who lack a thorough understanding of the ways computers and communications systems work. This has produced a guiding assumption about the need for "transparency" in the processing of personal data—an assumption that has led to some logical consequences about the fair information principles: about rights to access and correction, about rights to know about the existence of data-processing operations, about consent obligations and so on. Privacy and data-protection law is intended to increase trust in technologies and organizations by establishing some procedures by which the "lid can be taken off" the personal data-processing environment.[14]

To a large extent, however, this assumption was based on the vision of a set of relatively discrete personal information systems, established for legitimate and openly stated purposes, the contents of which could be registered and regulated. But where does one "personal information system" end and another begin in the more decentralized and networked "information highway" environment? Personal data is increasingly disclosed, matched, manipulated, and profiled in ways that may render any initial system or structure obsolete and meaningless.

The identification of common properties should not necessarily support a hypothesis of technological determinism. Technology "shapes the structure of the battle, but not every outcome."[15] It is still important to resist the conclusion that technological developments inexorably force states to adopt common policies. Policymakers and the institutions in which they operate have considerable autonomy to define the problem in particular ways and to make choices. It is apparent, however, that the "structure of the battle" is now observed at an international level. The difference in the dynamic is obvious from an examination of the attempts at policy harmonization within the European Union.

Policy Harmonization: The European Union's Data Protection Directive

The policy community that forged a consensus on the desirable content of data protection in the 1970s had, by the 1980s, expanded and become more institutionalized. The Annual Conference of the Data Protection Commissioners, begun as an informal "get-together" in the late 1970s, is now a major international convention. Two other international forums for cross-national debate have been the Council of Europe's Committee of Experts on Data Protection and the OECD's Information Computer Communications (ICCP) Division. The conferences on Privacy Laws and Business (in Britain) and Computers, Freedom and Privacy (in the United States) have been increasingly popular. In 1990, advocates of data protection and privacy formed Privacy International. Scholars of international relations would conclude that a cross-national epistemic community had coalesced by the end of the 1980s. Large numbers of individuals from government, business, academia, and consumer and public-interest groups now have stakes in national and international data-protection policy.[16]

The publication in September 1990 by the Commission of the European Communities of a Proposal for a Council Directive concerning the protection of individuals in relation to the processing of personal data appears, in retrospect, to have been a natural extension of earlier attempts to harmonize European data-protection policy.[17] In 1995, the European Union's Directive on the Protection of Personal Data and on the Free Movement of Such Data[18] finally emerged from the EU's complex and drawn-out legislative process. This could not have occurred in the absence of the network of experts and advocates. At the same time, its content was shaped by an increasingly transnational business community with interests somewhat opposed to the goals of the Directive.

Much has already been written about the European Union's Data Protection Directive.[19] Its provisions are complicated and reflect a lengthy process of bargaining in which a multitude of national and sectoral interests exerted influence. There are many derogations, qualifications, and alternatives available to member states, which may reduce important aspects of data protection to the "lowest common denominator."

The importance of this instrument stems from its status as a legally binding instrument. It therefore differs from both the "guidelines" of the OECD and the "convention" of the Council of Europe. Directives are designed to harmonize public policy throughout the European Union by expressing an agreed set of goals and principles, while granting member states some latitude (or subsidiarity) in deciding the actual ways in which those aims might be met. Thus, the goals of the Data Protection Directive are to "ensure a high level of protection for the privacy of individuals in all member states . . . and also to help ensure the free flow of information society services in the Single Market by fostering consumer confidence and minimizing differences between the Member States' rules."[20] This directive is one of several instruments that have been, or are being, debated, in the EU and which are designed to ensure that the free flow of capital and labor will be complemented by the free flow of information.

As European governments have until 1998 to bring their laws up to this new European standard, it is too early to predict the extent to which data-protection legislation will further be harmonized. Nevertheless, certain features of the Data Protection Directive suggest that the process of policy convergence begun in the 1970s will continue. Certainly the familiar set of "fair information principles" find further reinforcement. Beyond this, however, I would hypothesize a potential convergence of public policy in three further areas.

First, distinctions between "public" and "private" organizations were almost completely removed from the Data Protection Directive. Earlier drafts differentiated between the sectors in regard to the fair obtaining and processing of personal data, placing a greater onus on the private sector than on the public.[21] The final version largely avoids this distinction, perhaps reflecting the increasing difficulty of determining where the public sector ends and the private begins.

Second, many battles were fought over the question of the inclusion of "manual" data within the directive's scope. The British, whose 1984 Data Protection Act regulates only "automated" data processing, were strongly opposed to the regulation of non-automatically-stored personal data, arguing that the efficiency of storage and communication of personal data, which raised the data-protection issue in the first place, should continue to be the primary target of regulation. The final version reflects the opposite position—that "the scope of this protection must not in effect depend on the techniques used, otherwise this would create a serious risk of circumvention."[22] The directive covers, therefore, "structured manual filing systems" that form part of a "filing system." The British were given some extra time, beyond 1998, to bring manual filing systems under their law.[23] This issue was central to the decision of the British to abstain from the final vote in the Council of Ministers.

Third, the directive makes some strides in trying to harmonize the policy instruments through which data-protection principles are implemented. Article 18 [1], for instance, deals with the obligation of the data controller "to notify the supervisory authority . . . before carrying out any wholly or partly automatic processing operation or set of such operations." Moreover, "Member States shall provide that a register of processing operation notified in accordance with Article 18 shall be kept by the supervisory authority" (Article 21 [2]). This register is to include a range of information, including the identity of the controller, the purpose of the processing, the recipients of the data, and any proposed transfers to third countries. "Notification" seems to contemplate a form of registration similar to that within the UK Data Protection Act of 1984, with exemptions for more routine personal data and optional registration for manual records. It seems that the principle of "notification" can be seriously translated into practice only by means of a system of registration that many experts have regarded as excessively burdensome and expensive.

The Data Protection Directive also specifies the nature and function of a member state's "supervisory authority." Each country must assign one or more public authorities responsibility for monitoring the application of the national provisions adopted pursuant to the directive (Article 28 [1]). These authorities must act with "complete independence" and must be endowed with investigative powers, effective powers of intervention (especially before processing operations begin), powers to engage in legal proceedings, and powers to "hear claims" concerning the protection of rights and freedoms and regarding the "lawfulness of data processing" under the directive. In addition, member states are to consult the supervisory authorities when developing administrative measures with privacy implications. In total, these provisions require a greater range of powers and responsibilities than exist within most European data-protection regimes.

Thus, for the first time in an international agreement, we see an attempt to extend the process of policy convergence to the policy instruments. While there is much margin for maneuver, the Data Protection Directive begins the process of codifying a consensus on the most effective ways to implement data-protection law. This was anticipated in *Regulating Privacy* in the observation that "data protection officials will learn more and more about the approaches of their colleagues overseas and will try to draw lessons about the most effective responses to these common challenges."[24] The directive, however, signifies that the functions and powers of data-protection authorities will not simply be something that states can choose to fulfill in response to the emulation of overseas experience. Rather the policy instruments have become a matter for policy harmonization, and an obligation under European law.

Data Protection through Penetration: The External Impact of the EU's Data Protection Directive

The Data Protection Directive has had, and will continue to have, an impact on the data-protection policies of states that have not yet passed such legislation, including those outside the European Union. As of January 1996 only two EU countries (Greece and Italy) have yet to pass such a data-protection law; both are expected to do so in the near future. The directive is also influencing the policies of countries that are seeking membership of the EU; in 1992, for example, Hungary became the first country within Eastern Europe to pass legislation and establish a data-protection commissioner. Within Europe the directive has succeeded in increasing the degree of harmonization on the principles of data protection and begun to forge a consensus on the best means to implement those principles.

The pressure on non-EU countries stems principally from the stipulation in Article 25 that data transfers to a "third country" may take place only if that country ensures an "adequate level of protection." The "adequacy" of protection is to be assessed "in the light of all the circumstances surrounding a data transfer operation or set of data transfer operations." Consideration is to be given to the nature and purpose of the data, the "rules of law, both general and sectoral," and the "professional rules and security measures which are complied with" (Article 25 [1-2]). Article 26 lists a number of derogations, among which is the provision that data may be sent to countries with "inadequate" protection if the data controller enters into a contract that "adduces sufficient guarantees with respect to the protection of the privacy and fundamental rights and freedoms of individuals" (Article 26 [2]).

Where the European Commission decides that a third country does not ensure adequate protection, member states are "to take the measures necessary to prevent the transfer of data of the same type to the third country in question" (Article 25 [4]). Then the Commission "shall enter into negotiations with a view to remedying the situation" (Article 25 [5]). If the Commission finds an inadequate level of protection, member states are mandated, rather than simply permitted, to prohibit the transfer through what Paul Schwartz calls a "data embargo order."[25] This represents a stronger approach than that embodied within either the OECD Guidelines or the Council of Europe Convention. Even though each of these instruments contains a principle of "equivalence" (stronger than adequacy), neither one requires its signatories to block transfers of data to countries that cannot ensure equivalent protection. So, although the EU directive adopts a weaker standard, it embodies a stronger method of enforcement.[26]

The implications of Articles 25 and 26 have been debated at length by privacy experts. Conceivably, serious implementation of these provisions could have major economic consequences for credit-granting and financial institutions, for hotel and airline reservations systems, for the direct-marketing sector (including the list-rental business), for life and property insurance, and for any other sector that relies on the flow of personal data across borders.[27] Commentators are generally agreed that the Europeans are not going to tolerate the existence of "data havens"—jurisdictions in which data processing may take place in the absence of data-protection safeguards. The Data Protection Directive would be doomed to failure if multinational corporations could instantaneously transmit their processing offshore in order to avoid the transaction costs of having to abide by the stronger European rules.

On the other hand, there is disagreement over whether these provisions will actually be enforced in a coordinated way. The initial determination of "adequacy" will remain with the national data-protection agencies, who will still be implementing national laws that may diverge in some important respects.[28] There is also the danger that judgments about adequacy will be susceptible to the vagaries of the European policy process and are likely to be confused with the resolution of issues that have nothing to do with information privacy. Logrolling may therefore override the more predictable and rational pursuit of a common data-protection standard.[29] There is also confusion about whether "adequacy" will be judged against the principles of the directive or also against the methods of enforcement and oversight. These are crucial questions for businesses in North America that have developed "voluntary" codes of practice in the hope of preempting regulation.

The vagueness and unpredictability of Articles 25 and 26 have caused variable amounts of concern in countries outside the European Union. Nowhere has the Data Protection Directive been the sole reason for another country's passing a data-protection law. On the other hand, it has certainly been one important influence. The early draft of the directive was not far from the minds of those who drafted New Zealand's 1993 Privacy Act, for example.[30] It was continually referenced in the report that led to a data-protection law in Hong Kong.[31] In 1993, Quebec became the first jurisdiction in North America to enact private sector data protection.[32] The 1993 Quebec law was based on the earlier EC Draft Directive[33] and was drafted explicitly to protect Quebec business from the possible blockage of data transfers from Europe.[34] Quebec is now the only jurisdiction in North America that can unequivocally claim an "adequate level of protection."

The EU's Data Protection Directive has also had an impact on the slow emergence of data protection as an item on Canada's federal agenda. In 1992, the Canadian Standards Association—motivated by the EU Directive and by the fear that "EC member countries may be reluctant to share personal data with Canadian businesses, potentially creating a trading block of immense consequence"[35]—began negotiating a Model Code for the Protection of Personal Information. This was followed in 1995 by a recommendation from the Canadian Information Highway Advisory Council that the federal government "create a level playing field for the protection of personal information on the Information Highway by developing and implementing a flexible legislative framework for both public and private sectors."[36] This recommendation was accepted by the federal government in May 1996. If legislation emerges, it will undoubtedly be based on the new CSA standard.[37] A measure of the seriousness with which the issue is now taken is the recent call for national legislation by the Canadian Direct Marketing Association.[38] The economic implications of a set of international standards for the protection or personal information which Canada could not claim to satisfy have never been far from the background considerations of each of these agencies.

There is no doubt, therefore, that the EU's 1995 Data Protection Directive now constitutes the rules of the road for the increasingly global character of data-processing operations.[39] Its effect on countries outside the EU is principally a penetrative one. The increasing global interdependence means possible consequences for those businesses that rely upon the unimpeded flow of personal information, and which cannot claim to protect the data of consumers, clients and employees in ways that match the EU standard. Economic motivations have therefore been increasingly prominent in the data-protection debates of the 1990s.

The Desire to Be in the "Data-Protection Club": The Emulation of the Majority

The penetrative process assumes a relative lack of state autonomy. In an interdependent world, the policy efforts of the Europeans carry externalities that force other countries to pursue policies that they would otherwise oppose or avoid. The alternative is to bear the costs of maintaining a different public policy. In addition, the general pressures to conform have increased as more and more countries have joined the "data-protection club." There is an increasing perception that adequate privacy protection is a necessary condition for being on the global information highway.

The commercial motivations for certain countries have been significant.[40] There is also, however, a psychological concern for being within a minority. Privacy and data protection are increasingly seen as one procedural requirement of a modern democratic state. In the 1970s and the 1980s, when omnibus data-protection policy was confined to Western European societies, it was possible for some commentators in Canada, Australia, and the United States to argue plausibly that this legislation was a feature of a continental (civil) legal system, and that the Anglo-American system dictated a less regulatory regime that placed more responsibility on the individual citizen to demonstrate damage and make a claim through the courts. Those arguments have sounded increasingly specious since the United Kingdom and New Zealand passed data-protection acts (in 1984 and 1993, respectively). It has also become difficult for North Americans to claim that Canada and the United States should have weaker privacy safeguards than Spain, Portugal, the former East Germany, and Hungary, countries that in recent memory have been governed by authoritarian regimes.

The EU directive, therefore, is not only an instrument of pressure and economic interdependence. It is also conceivably a legal framework for others to emulate. It represents the "most modern international consensus on the desirable content of data-protection rights," and it "may be a valuable model for countries currently without data protection laws."[41] The emulation of overseas public policy was a significant force in the early years of this policy, when there was little literature, much insecurity, and a natural desire to emulate the few pioneers. Today, data protection is not an innovation. It is an expectation that works in favor of the majority of countries, producing a desire to "keep up with Joneses" in every country except the United States.

The Divergence of Policy Experience

The pressures for further policy convergence seem strong, indeed inexorable. At the level of policy content, and to a certain extent at the level of policy instruments, data-protection policies are more alike today than they were in the mid 1980s. More countries have data-protection laws, and those laws look more similar. There are, however, some countervailing forces that suggest a continuing divergence.

American Exceptionalism

At the end of 1996, of 24 OECD countries, only six have failed to enact a comprehensive privacy law that applies the fair information principles to all organizations that process personal data: the United States, Canada, Australia, Japan, Greece, and Turkey. Greece will be forced to produce legislation based on the EU directive by 1998, and bills have been debated in the Greek legislature. A data-protection bill has also been prepared and debated in Turkey. The Australian government has just decided to "develop an effective, comprehensive scheme to protect individual privacy while avoiding unnecessary burdens on business and the community."[42] Canada's and Japan's governments have passed privacy legislation for their governmental agencies and have encouraged self-regulation in the private sector. The Canadian federal government has recently been advised by the Information Highway Advisory Council and others to create a more consistent and enforceable set of "rules of the road" for the information highway by developing and implementing a flexible legislative framework for the protection of personal information in the public and private sector.[43] In May 1996, the federal Minister of Industry announced the government's intention to "bring forward proposals for a legislative framework governing the protection of personal data in the private sector."[44]

That leaves the biggest and most powerful advanced capitalist country, the United States. Here, as in Canada and Australia, the public sector is largely regulated through the 1974 Privacy Act and equivalent statutes at the state level. The private sector is regulated through an incomplete patchwork of federal and state provisions that oblige organizations to adhere to fair information practices.[45] But there is no oversight agency for privacy in the United States, and some would argue that there are few effective remedies.[46] The approach to making privacy policy in the United States is reactive rather than anticipatory, incremental rather than comprehensive, and fragmented rather than coherent.[47] There may be a lot of laws, but there is not much protection.

There is little indication that this state of affairs is likely to change. The formulation of privacy policy has always been susceptible to the unpredictable vagaries of a fragmented political system in which a variety of interests are continuously in play.[48] It is very easy to prevent legislation in the United States through the many "veto points" within the complicated legislative process. In the best of political climates, any attempt to pass a comprehensive data-protection law overseen by a privacy agency would undoubtedly be challenged by an army of well-paid corporate lobbyists. Under the current anti-regulatory and fiscally conservative climate of the Congress, such an outcome would be inconceivable. American privacy advocates have also had other more defensive battles to fight, especially against the "Clipper chip" and the associated FBI wiretap legislation.

Thus, by the end of the 1990s there could very well be data-protection legislation in every advanced industrial country (and some others besides) with the exception of the United States. Will this mean that the United States will be one vast "data haven"? Will the Europeans (and others) have the political will and administrative means to enforce restrictions on the flow of data to the United States? American multinational corporations have so far proceeded under the assumption that individual contracts could suffice to convince the Europeans that their data-processing practices are "adequate." If that is true, then the policy convergence observed in the rest of the world would not extend to the American political system. It may extend to a few of the most sensitive multinational players, but the United States as a whole would remain the huge exception. And this exceptionalism will no doubt continue to be rationalized by Americans in terms of their different political system, more individualistic political culture, and more reactive policy style.[49]

Privacy Codes and Standards
Some analysts have observed a discernible shift in the mid 1980s to legislation that displayed a greater sensitivity to the particular information-processing practices and technological conditions of different sectors. This move toward sectoral regulation is reflected in the "third-generation" data-protection statutes, of which the 1988 Netherlands and 1993 New Zealand laws are the most notable.

The mechanism to implement a greater flexibility is typically the "code of practice." In countries with data-protection legislation, these codes are intended to translate the broader legislative requirements into practical advice and a more pragmatic interpretation of the legislative wording. In New Zealand and the Netherlands, such codes are formally endorsed by the respective data-protection agencies and then have the force of law. In the United Kingdom, however, codes of practice are encouraged by the UK Data Protection Registrar as instruments to propagate "good computing practice" but they do not have the force of law.[50]

Sectoral codes of practice are becoming increasingly popular, even in countries whose laws do not explicitly encourage their use. The EU directive supports their development and implementation.[51] They are flexible instruments, the development of which can increase the understanding within organizations of personal data collected, stored, and processed. Codes can reduce suspicions about the improper collection and use of personal data. Their development can allow a thorough review of practices for the processing of personal data.[52]

Codes of practice also, however, institutionalize a certain divergence in policy implementation. The adaptation of a set of broad legal guidelines into more practically specific advice for different sectors entails an inherent divergence of approach. Codes are based on the assumption that the necessary rules for the health sector, the financial sector, the direct-marketing industry, the credit-reporting business, and so on must be somewhat different. It is a continual challenge for those data-protection agencies that negotiate codes to ensure that the search for relevance and precision does not lead to an unnecessary weakening or dissipation of the legislative standard.

In the United States and Canada, privacy codes have generally been developed in conformity with the OECD Guidelines of 1981. These "voluntary" codes are now quite numerous but display considerable variability in terms of scope and extent of enforcement. Some have been developed speedily in order to provide a symbolic response to complaints or to avoid a more onerous regulatory regime. Others have been developed with relative care and on the basis of more careful internal audit and external consultation. Some embody external mechanisms for complaints resolution and oversight; others are merely statements of good intention.[53]

The diversity of codes of practice in Canada was one factor that prompted the Canadian Standards Association to negotiate a Model

Code for the Protection of Personal Information with business, government, and consumer groups. Since 1992, a committee of stakeholders has been updating and revising the OECD Guidelines with reference to the Quebec legislation and the emerging EU directive. The CSA Model Code was passed without any dissenting vote on September 20, 1995, and was subsequently approved by the Standards Council of Canada. The privacy code is then to be adopted by different sectors, adapted to their specific circumstances, and used to promote fair information principles in both public and private organizations.

The CSA Model Code is a rather different instrument from those developed by companies or trade associations. It is a standard, and it has the potential to operate in the same way as many other quality-assurance standards (such as the increasingly popular ISO 9000 series). Organizations would be pressed by government, clients, and/or consumers to demonstrate that they implement fair information practices, would adopt the standard, and would then be obliged to register with an accredited certification or registration body. Scrutiny of internal operational manuals and/or on-site auditing could be a prerequisite for maintaining a registration.[54]

The possible future integration of data-protection policy into the institutional environment of standard setting could presage a further divergence of public policy. If the CSA standard were to be adopted in Canada and perhaps in other countries,[55] some organizations would be forced through commercial, consumer, and government pressure to demonstrate high privacy standards—perhaps even higher than their counterparts in countries with data-protection legislation. However, the spread of privacy policy would be piecemeal, incremental, and unpredictable. A standards approach is based on the assumption that higher levels of privacy awareness are required in some organizations than in others, and that over time these higher standards would be forced through marketplace pressure. A standards approach to privacy protection would force a greater convergence of organizational practice only after considerable diffusion of the standard and the attainment of a threshold beyond which other organizations faced considerable costs of non-compliance. To the extent that such a threshold would not be reached, a privacy standard would reinforce a divergence of practice between and even within sectors.

Privacy-Enhancing Technologies

There is now no question that information technologies can be a friend to privacy as well as a threat. They can be designed with zero collection of personal information as the default setting. They can be designed to delete and forget personal data. Encryption can be incorporated into software and embedded in memory chips in computers or smart cards. Public-key cryptography can be used to verify transactions without betraying individual identity and without leaving a trail of "transactional data."[56] I do not need to analyze the complexities of public-key cryptography and digital signatures and outline their many possible applications to argue that the spread of these technologies will be variable and unpredictable, and that several factors will determine the success of their application.

In the market context, data on personal transactions has economic value. Applications based on public-key encryption deny organizations the ability to develop more sophisticated analyses and profiles of various market segments. The willingness to make that tradeoff will then be governed by the salience of privacy for those consumers at that time in that market relationship. Our public opinion surveys suggest that this level of concern is highly variable across demographic groups.[57]

The widespread application of privacy-enhancing technologies will also require a massive educational effort. The theoretical basis of these technologies is sound and coherent, but the argument to be made at the level of organizational and public education is difficult to convey and somewhat counterintuitive. Those societies with data-protection agencies that have an educational mandate and that pursue that mandate with vigor will be in a better position to place technologies of privacy on the agenda.

Privacy-enhancing technologies cannot be a substitute for public policy. They may be influenced by the slow development of standards. In the long term, their application will be influenced by governmental attempts to establish encryption policies for law-enforcement purposes and for the electronic delivery of government services. Some governments will no doubt prefer "key-escrow" policies and standards (such as the heavily criticized "Clipper chip" proposal in the United States). Others may prefer to influence the development of "public-key infrastructures," as has been recommended to the Canadian federal government.[58] Overall encryption policy, outstanding legal questions (such as the legal status of "digital signatures"), fluctuating market pressures, and rapidly advancing technological developments will ensure considerable tensions within and between countries.

The application of these technologies will then be governed by the complex interplay of market forces, consumer pressure and education, and governmental sponsorship. The result, at least in the foreseeable future, will be considerable variation in privacy and security standards within and between sectors. There is little evidence at the moment that these tensions will be resolved in ways that will advance policy convergence.

Conclusion: The Limits to Harmonization

In view of the above-mentioned distinction among policy content, policy instruments, and policy outcomes, let us now identify the trends toward convergence and divergence across each of these dimensions of data-protection policy.

The European Union's Data Protection Directive will undoubtedly extend the process of policy convergence at the level of policy content (the agreed legislative principles of a data-protection policy). Countries within Europe have until 1998 to amend their respective laws. Countries outside Europe will increasingly use the EU Directive as a model in devising or updating their laws. The EU Directive will not only be an instrument for harmonization within Europe; it will have a more coercive effect on countries outside. It is conceivable that by the end of the century every advanced industrial democracy with the exception of the United States will have granted its citizens enforceable legal rights over personal information collected by any organization.

The EU Directive will also push a greater conformity at the level of policy instruments (the agencies of oversight, enforcement, and implementation). The specification of the powers that should be granted to a "supervisory authority" will probably create greater conformity in the operations of European data-protection agencies. Moreover, the principle of independent oversight may also be regarded as a test of "adequacy" for the transmission of personal data outside Europe. In this case, the "voluntary" codes of practice administered through individual companies or trade associations in North America may not suffice.[59]

At the level of policy outcome, however, evidence for policy convergence is far more difficult to trace. How then do we know that these increasingly similar data-protection laws, overseen by increasingly similar data-protection agencies, are having similar effects? Generally, we do not. Few data-protection agencies have the time and resources to conduct regular privacy audits. None can make unassailable claims that the implementation of its laws has produced common and higher standards of privacy protection. We can make plausible inferences, but the enactment of law does not necessarily change behavior. The translation of fair information principles into fair information practice is a slow, halting, and frustrating process. The change requires more than legal fiat and regulatory power.

At root, the problem may be conceptual rather than political, technological, or economic. The EU Data Protection Directive aims for a "high and common level of protection." It appears that the debates in the EU Commission and Council of Ministers have centered on whether there should be high or minimal levels of protection, and high or lower levels of harmonization.[60] Nowhere, however, does the directive define how that level is to be observed. Data-protection policy aims to satisfy a highly elusive and subjective value: the right to privacy. It dictates changes in the processing of a highly ephemeral resource with variable value: personal information. Rhetoric about the "level" of protection reflects a fundamental misunderstanding about the properties of this public policy.

Data protection is not like environmental protection, in which states might agree on the desirable "level" of toxins in rivers and have a relatively clear and common understanding of what that "level" means. For data protection we can compare the "black letter of the law," we can observe indicators of the scope of law (manual vs. automated data, public vs. private, etc.), and we can compare and contrast the functions and powers of the policy instruments. But it is fallacious to make inferences about the "level of protection" from the observation of these crude indicators. Any attempt to establish evaluative criteria for assessing performance is fraught with the central difficulty that the goals of data protection are not self-defining.[61]

What is needed is a more holistic perspective that sees data protection as a process that involves a wide network of actors (data users, data subjects, and regulators) all engaged in the co-production of data protection. The successful implementation of data protection requires a shift in organizational culture and citizen behavior. Data protection is a learning exercise that involves a mutual process of education and mediation from the bottom up as much as it involves regulatory command from the top down.[62] It is a public policy singularly unsuitable to the traditional techniques of "rational" policy analysis, the unambiguous definition of goals, and the measurement of goal attainment over time, sectors, countries, technologies, and so on. There are, then, clear limits to the evaluation of policy success, and thus to the observation of policy convergence. Rhetoric to the contrary, we may well have confronted those limits.

Notes

1. Colin J. Bennett, *Regulating Privacy: Data Protection and Public Policy in Europe and the United States* (Cornell University Press, 1992).

2. Organization for Economic Cooperation and Development, Guidelines on the Protection of Privacy and Transborder Flows of Personal Data, 1981.

3. Council of Europe, Convention for the Protection of Individuals with Regard to the Automatic Processing of Personal Data, 1981.

4. Bennett, *Regulating Privacy*, p. 222. It should be noted that my analysis concentrated principally on the public sector. For these purposes, Canada and the US were considered innovators, even though neither country has yet to pass comprehensive data-protection legislation that also covers the private sector.

5. See Colin J. Bennett, Implementing Privacy Codes of Practice: A Report to the Canadian Standards Association, Report PLUS 8830, Canadian Standards Association, 1995.

6. David H. Flaherty, *Protecting Privacy in Surveillance Societies* (University of North Carolina Press, 1989).

7. Christopher C. Hood, *The Tools of Government* (Chatham House, 1986).

8. Besides my work, other contributions from political science include Priscilla M. Regan, *Legislating Privacy: Technology, Social Values and Public Policy* (University of North Carolina Press, 1995), and Charles D. Raab, Data protection in Britain: Governance and learning, *Governance* 6 (1993): 43–66. For an analysis of political science's contribution to the study of privacy see Colin J. Bennett, The Political Economy of Privacy: A Review of the Literature, written for Human Genome Project, Center for Social And Legal Research, Hackensack, N.J., 1995.

9. Colin J. Bennett, Review article: What is policy convergence and what causes it? *British Journal of Political Science* 21 (1991): 215–233.

10. This globalization is a prominent theme in every analysis of the information highway. For Canada, see Industry Canada, Privacy and the Canadian Information Highway (Industry Canada, 1994); Information Highway Advisory Council, Connection, Community, Content: The Challenge of the Information Highway (Minister of Supply and Services, Canada, 1995). For the US, see United States Information Infrastructure Task Force, Privacy and the National Information Infrastructure: Principles for Providing and Using Personal Information, Final Version, June 6, 1996, Privacy Working Group, IITF Information Policy Committee.

11. Bennett, *Regulating Privacy*, p. 121.

12. Philip E. Agre, Surveillance and capture: Two models of privacy, *Information Society* 10 (1994): 101–127.

13. For instance, instruments such as the OECD Guidelines differentiate the collection, the use, and the disclosure of personal information. The EU Data Protection Directive recognizes the difficulty of sustaining this distinction in practice and speaks more generally about "personal data processing, meaning "any operation or set of operations which is performed upon personal data" (Article 2[b]).

14. Bennett, *Regulating Privacy*, p. 121.

15. Ithiel de Sola Pool, *Technologies of Freedom* (Harvard University Press, 1983), p. 251.

16. See Colin J. Bennett, The International Regulation of Personal Data: From Epistemic Community to Policy Sector, presented at 1992 meeting of Canadian Political Science Association, Prince Edward Island.

17. European Commission, Proposal for a Council Directive Concerning the Protection of Individuals in Relation to the Processing of Personal Information, COM (90) 314 Final-SYN 287, 1990.

18. European Union, Directive 95/46/EC of the European Parliament and of the Council on the protection of individuals with regard to the processing of personal data and on the free movement of such data, OJ No. L281, 1995.

19. See Graham Greenleaf, The 1995 EU directive on data protection—An overview, *International Privacy Bulletin* 3 (1995), no. 2: 1–21; Paul M. Schwartz, European data protection law, *Iowa Law Review* 80 (1995), no. 3: 471–496; Spiros Simitis, From the market to the polis: European data protection law and restrictions on international data flows, *Iowa Law Review* 80 (1995), no.): 445–469.

20. Mario Monti, Council Definitively Adopts Directive on Protection of Personal Data, press release IP/95/822, European Commission, July 25, 1995.

21. European Commission, Proposal for a Council Directive, September 1990.

22. EU Data Protection Directive, Recital no. 27.

23. Chris Pounder and Freddy Kosten, The data protection directives, *Data Protection News* 21 (1995), spring: 2–38.

24. Bennett, *Regulating Privacy*, p. 251.

25. Schwartz, European Data Protection Law, p. 488.

26. Greenleaf, The 1995 EU Directive, p. 16.

27. Charles D. Raab and Colin J. Bennett, Protecting privacy across borders: European policies and prospects, *Public Administration* 72 (1994): 95–112.

28. Schwartz, European data protection law, p. 495.

29. Pounder and Kosten, Data protection directives, p. 33.

30. Interviews, Auckland and Wellington, September 1992.

31. Law Reform Commission of Hong Kong, Report on Reform of the Law Relating to the Protection of Personal Data, August 1994.

32. Bill 68, an act respecting the protection of personal information in the private sector (Quebec Official Publisher, 1993).

33. European Commission, Amended Proposal for a Council Directive on the Protection of Individuals with Regard to the Processing of Personal Data and on the Free Movement of Such Data, COM (92) 422 Final-SYN 287. Brussels, 1992.

34. Paul-Andre Comeau, The Protection of Personal Information in the Private Sector: An Important Step Forward by Quebec's National Assembly, address to the Sixth Annual Conference of Privacy Laws and Business, 1993.

35. Privacy code a must for global economy, *Focus* (Canadian Standards Association), spring 1992.

36. Information Highway Advisory Council, Connection, Community, Content: The Challenge of the Information Highway (Ottawa Minister of Supply and Services, 1995), p. 141.

37. Canadian Standards Association, Model Code for the Protection of Personal Information, CAN/CSA-Q830, 1995.

38. Canadian Direct Marketing Association, Direct Marketers Call for National Privacy Legislation, news release, October 3, 1995.

39. Joel Reidenberg, Rules of the road for global electronic highways: Merging the trade and technical paradigms, *Harvard Journal of Law and Technology* 6 (1993): 287–305.

40. The commercial motivations were especially strong within Britain. See Bennett, *Regulating Privacy*, pp. 141–143.

41. Greenleaf, 1995 EU Directive, p. 1.

42. Government Moves to Protect Privacy, Australian government news release, December 10, 1995.

43. Information Highway Advisory Council, Connection, Community, Content, p. 141.

44. Building the Information Society: *Moving Canada into the 21st Century* (Industry Canada, 1996) (http://info.ic.gc.ca/info-highway/ih.html).

45. See Robert Ellis Smith, *Compilation of State and Federal Privacy Laws* (Privacy Journal, 1992).

46. See Marc Rotenberg, *Privacy Law in the United States: Failing to Make the Grade* (Computer Professionals for Social Responsibility, 1991).

47. Robert M. Gellman, Fragmented, incomplete and discontinuous: The failure of federal privacy regulatory proposals and institutions, *Software Law Journal 6* (1993): 199–238.

48. See Priscilla M. Regan, *Legislating Privacy: Technology, Social Values, and Public Policy* (University of North Carolina Press, 1995).

49. H. Jeff Smith (*Managing Privacy: Information Technology and Corporate America*, University of North Carolina Press, 1994) has characterized the prevailing corporate approach to privacy protection as "drift and react."

50. For a discussion of the role of codes of practice within data-protection law see Bennett, Implementing Privacy Codes of Practice, pp. 43–49.

51. EU Data Protection Directive, Article 27.

52. See Peter Hustinx, The Use and Impact of Codes of Conduct in the Netherlands, paper presented to Sixteenth International Conference on Data Protection, The Hague, 1994; Alan F. Westin, Managing consumer privacy issues: a checklist, *Transnational Data and Communication Report*, July-August 1991: 35–37.

53. For an analysis of privacy codes in Canada see Bennett, Implementing Privacy Codes of Practice. For a recent compilation of US codes see *Privacy and American Business, Handbook of Company Privacy Codes* (Bell Atlantic, 1994).

54. The CSA also offers a certification and registration service to Q830.

55. Discussions have already begun in the International Standards Organization about the feasibility of a separate ISO privacy standard based on the Canadian model.

56. For a concise and approachable explanation of "technologies of privacy" see Ann Cavoukian and Don Tapscott, *Who Knows: Safeguarding Your Privacy in a Networked World* (Random House Canada, 1995), pp. 132–142.

57. The most recent report in Canada (*Surveying Boundaries*, Public Interest Advocacy Center, 1995) bears out the different levels of concern for privacy depending on the tradeoff.

58. See Connection, Community, Content, pp. 145–146.

59. Greenleaf, 1995 EU Directive, p. 17.

60. EU heads of government support adoption of data protection draft directive by end of 1994, *Privacy Laws and Business*, September 1994.

61. Charles D. Raab and Colin J. Bennett, Taking the measure of privacy: Can data protection be evaluated? *International Review of Administrative Sciences 62* (1996), December: 95–112.

62. Charles D. Raab, Data protection in Britain: Governance and learning, *Governance 6* (1993), no. 1: 43–66.

4

Privacy-Enhancing Technologies: Typology, Critique, Vision

Herbert Burkert

The term *privacy-enhancing technologies* (PETs) refers to technical and organizational concepts that aim at protecting personal identity. These concepts usually involve encryption in the form of (e.g.) digital signatures, blind signatures, or digital pseudonyms.[1]

PETs have to be set apart from data-security technologies. It is one of the merits of the discussion on PETs that the concept of data security has been reclarified as to its limitations with regard to *privacy protection*. Data-security measures seek to render data processing safe regardless of the legitimacy of processing. Data security is a necessary but not a sufficient condition for privacy protection. PETs, on the other hand, seek to eliminate the use of personal data altogether or to give direct control over revelation of personal information to the person concerned. PETs are therefore closer to the social goals of privacy protection. How close will be one of the issues of this chapter.

PET concepts start from the observation that personal information is being accumulated in interactions between persons or through the observation of such interactions. This starting point provides the structure for a typology. If we see such interactions as actions between subjects relating to objects within systems, we can differentiate four types of PET concepts:

- subject-oriented concepts
- object-oriented concepts
- action-oriented concepts
- system-oriented concepts.

Not all of these types have yet been fully discussed or implemented. None of these concepts is being implemented in its pure form. This typology should therefore be used only as a heuristic instrument to sort out the main characteristics of such systems.

Subject-Oriented Concepts

Subject-oriented concepts seek to eliminate or substantially reduce the capability to personally identify the acting subject (or interacting subjects) in transactions or in their relationship to existing data.

Within this concept, this aim may be achieved by implementing proxies: Individuals may be given identifiers. These identifiers may be untraceable with regard to a particular transaction or a set of transactions or a set of data relating to the subject(s). These identifiers may remain stable or they may be generated specifically for each transaction. The generation of identifiers may follow a preset rule (making sure that this rule does not compromise the identity), or identifiers may be generated at random. These identifiers may or may not be retraceable. If they are retraceable, such connections may be organized in a way that ensures that the rules for such connections are being followed. Rule keeping for such tracing processes may be given to organizations or individuals ("trusted third parties"), or the links may be under the control of a technical device that needs special procedures to be followed in order to establish such links.

For example, to implement a credit card system as such a PET system we could take the name of the cardholder and the link to the credit card number from the system and put them into a "data safe" that makes such information accessible only with the consent of the individual. For this purpose this information could be encrypted, using a public-key encryption system, in such a way that both the key of the bank and that of the account holder are needed to make such information readable. We could do the same thing for the information on the bank account connected to that card. This would allow us to use a "credit card" to do business—at least with regard to non-personalized goods and services—without the other parties' knowing about our identity.

Object-Oriented Concepts

Object-oriented concepts build on the observation that transactions often involve exchange or barter and the observation that an object given in exchange carries traces (not unlike fingerprints) that allow the identification of the exchanging subjects. Within these concepts, the aim then is to free the exchanged object from all such traces without eliminating the object itself. The most common object of exchange is, of course, payment in a denomination. The main example of payment with reduced identification capabilities is money in the form of cash. Object-oriented concepts seek to construct electronic equivalents to cash or barter objects that do not allow the object to be connected to the subject handling it. An electronic impulse coming from an anonymous charge card paid for in cash would constitute such an electronic equivalent. Or, rather than separate the identifiers from the card and the account number, a bank might give you a series of numbers, each of them worth a fixed amount of money, which you could use as freely as cash.

Transaction-Oriented Concepts

Transaction-oriented concepts—concepts which I have not yet seen discussed extensively, at least not in the context of PETs—seek to address the trace the transaction process is leaving behind without directly addressing the objects being exchanged. (Tracing processes may take place at banks when over-the-counter cash dealings are videotaped, or when such cash payments are made against receipts.)

Transaction-oriented concepts would provide for the automated destruction of records. For example, transaction records might be automatically destroyed after a preset time. In this context one could make use of technical devices like combinations of data and small programs that are currently being discussed in the context of the protection of intellectual property rights (e.g., electronic records or software programs that destroy themselves if no payment mechanism is initiated within a given time).

System-Oriented Concepts

Finally, we might envisage concepts that seek to integrate some or all of the elements described, thereby creating zones of interaction where the identity of the subjects is being hidden, where the objects bear no traces of those handling them, and where no record of the interaction is created or maintained.

Current social equivalents that come to mind for such systems are Catholic confession (provided the parish is large enough) and anonymous interactions with crisis intervention centers. In an electronic context, two parties communicating via e-mail, each of them using a service that anonymizes the communication and destroys the traces of this communication at the end, might be the closest equivalent.

Attempt at a Critique

The Achievements of PET Design

It has been made clear from the beginning by those advocating PETs that the first question to be asked in PET design is whether personal information is needed at all. The availability of PETs creates a burden of legitimization on those who want to have personal information in their information systems.

A further achievement of PETs is that they take pressure off the consent principle. Too often, designers of systems intended to handle personal information have sought to avoid problems caused by handling such information by seeking the consent of the subjects. Rather than put effort into avoiding the use of personal information or into adjusting the system to specific legal requirements, they simply seek to get the subject's consent for whatever they wish to do with personal information. The "consent approach" is too often applied without sufficiently analyzing the range of "true" choices available to the data subject. Rather, the data subject is asked to choose between giving consent and losing advantages, privileges, rights, or benefits, some of which may be essential to the subject in a given situation. Consent in the context of privacy protection also means informed consent; the degree of information provided, however, tends to vary widely, with the data subjects at a disadvantage because they cannot know how much they should know without fully understanding the system and its interconnections with other systems. With the wider availability of PET designs, one could avoid from the beginning situations in which consent would have to be requested and situations in which "consent" would be collected in such an imperfect manner.

On the other hand, one should ensure that the PET approach does not set in before the consent principle has been examined. PET systems do not replace the consent requirement. Their use will most likely occur in areas where no choice exists, where no real consent could be obtained in the first place because the person concerned is dependent on a particular interaction,[2] and where all efforts would have to be made to design the information system to be as risk free as possible.

PETs as a concept have brought the social "soft" concern of how to reduce personal data down to the level of "hard" system-design considerations.

Within this context, PETs first of all bring to our attention the privacy-protection principle, explicitly enshrined in all data-protection laws, that personal data should only be collected if *necessary*.[3] Then, with the ability to point to the specifications of PETs as *system-design* specifications, privacy-protection proactivists are able to talk to system designers on the same "language" level, translating their values and concerns into system-design language.[4] With this step, arguments of privacy protection can no longer be brushed aside as being "untranslatable" or being outside the competence (and obligations) of design thinking and the "engineering approach."

Addressing privacy on the design level is an essential precondition to addressing privacy on the *political* level: In technology-oriented societies the ability to "do something" is an important element in strategies of legitimization. With the existence of PETs, whatever (personal-information) systems are regarded as being politically desirable now have to be justified on the basis of the extent to which have taken into account the technical options that are already available on the *technical* level to minimize the occurrence of personal information. In consequence, political intentions to create sources of personal information or to increase their availability can no longer hide behind technical necessities, design difficulties, or habits of systems thinking, but clearly have to state their *political desire* to make personal information available as identifiable information. Consequently the political responsibility for any risks resulting from such an availability can be designated more clearly.

The Limitations of PET Design

With these achievements, however, we also have to clearly identify or restate the limitations of this design approach so as not to lose what we have achieved in the implementation debates. Some of these limitations are inherent in the designs and are generally understood to be so by PET designers. Other limitations do not apply to all PET concepts alike and could be remedied by modifying the design. Some of the limitations are external. Most of these limitations have already been seen by those advocating PETs. I will discuss them in detail here.

Internal limitations

I see four internal limitations of PET design: the one-directional perspective, the problem of "identifying information," the problem the "systems perspective" poses in general, and the underlying technical assumption.

The one-directional perception

Some of the subject-oriented PET concepts build on the perception that within an interaction there is usually one party who merits to be guarded against his counterpart or an observing third party. The subject warranting protection will usually be shielded in a manner that allows that subject to identify his counterpart. I call this approach "one-directional." These concepts implement some sort of one-directional anonymity. The decision which side in the interaction warrants protection, however, is a normative and not a technical one. Following a normative concept such as "power has to be balanced," we might perceive individual consumers as needing protection against economically powerful providers of goods and services. Under this assumption, we would focus our attention on developing services in which the consumers can choose whether and to what extent they wish to identify themselves, insofar as this is feasible. (They still might want to receive what they have ordered anonymously.) We would not allow such devices for the seller. But how can we be sure that we have perceived the power relationship adequately? There might well be situations—for example, in the context of electronic data interchange—in which the buyer is a potent organization and the seller is a small company with a legitimate interest in being protected against the buyer's economic power.

This example has two implications: We should not forget that PETs—and in this they do resemble data-security designs—essentially remain technical: They *follow* the normative decision.[5] This normative decision still has to be made; it cannot be replaced by the PET concept as such. We have to *decide*—if taking such a one-directional approach—which of the parties merits such protection. We must therefore be careful, at least in some of their implementations, when addressing these technologies simply as "privacy-enhancing technologies" when in fact we are faced with *technical* instruments that protect identities. Whether this protection is indeed enhancing privacy (a social value) is another matter. The discussion of PETs has helped to show the limits of data-security concepts insofar as data security is value free (we might make the wrong sort of communications secure). We should therefore be careful not repeat such mistakes with PETs: they may indeed enhance privacy as we cherish it, but they might also enhance unwanted secrecy.

In political environments (e.g., that of the United States) that are less formed by a "freedom of information" culture, we might envisage the use of PETs by administrations that intend to protect the identities of persons who have to operate in hostile environments. Administrators would demand to be able to anonymize their participation in administrative decisions or actions by electronic signature processes, their personal identity to be revealed only under certain material grounds and with certain procedural precautions (e.g., only if a judge has ordered such identification). Number badges for policemen (instead of name badges) would be an example for such a system in a solely paper-based environment with similar advantages and disadvantages. Indeed, PETs are still being used to reinforce principles of administrative secrecy in Europe. PETs do not release us from addressing the need to balance privacy interests against transparency interests by making a *value* judgment.

In implementing PETs careful attention has thus to be paid to the normative question in order to handle the inherent "dual-use" problem of these technologies: PETs may make it possible to *maintain* a given distribution of organizational power rather than to empower individuals in their dealings with such organizational power and thus enhancing the defense of their rights and freedoms and privacy.

Identifying information
Some PET concepts rely on the capability to differentiate information that relates to a person from "identifying information" and on keeping these two types of information separate. As in our example above, we might separate and keep the name and address of a card holder (identifying information) separate from the credit card number. Legally, however, that

credit card number might still be personal information. Directive 95/46/EC of the European Parliament and of the Council of October 24, 1995, on the protection of individuals with regard to the processing of personal data and on the free movement of such data,[6] defines "personal information" as "any information relating to an identified or identifiable natural person ('data subject'); an identifiable person is one who can be identified, directly or indirectly, in particular by reference to an identification number or to one or more factors specific to his physical, physiological mental, economic, cultural or social identity. . . ."[7] The problem then lies, of course, in the term "indirectly": What sort of barrier is necessary to regard the possibility to link the credit card number to a person as at least "indirect"? Problems of this sort, however, divert us from a more fundamental matter: we might be faced with identifying information that is there without our noticing it, or additional information from other systems or contexts might be brought in, suddenly allowing identification. The capability to identify persons behind anonymous information depends on the purpose of our identification, on the "hidden information" connected to this anonymous information, and on the extent of "additional knowledge" that may be provided by other information systems.[8] All these elements, however, are difficult to predict at the stage of systems design. Having designed a particular PET system does not, in itself, guarantee that this system will be used only for the particular purpose it was intended for; we may not be sufficiently aware of hidden information such a system carries with it; we cannot be reasonably sure or predict which other non-PET systems are or will be available to provide "additional knowledge" that allows us to decipher anonymity, thus limiting the effect of the PET elements.

Let us assume that we establish a system in which all the members of a certain minority within a certain geographic unit are listed. This register is being run as a "subject-oriented" PET system; that is, the names and the addresses are kept separate from other information that is also kept by this system (e.g., sex, age, occupation, and income range). Let us further assume that the purpose of this system is to provide special services to this minority, such as planning schools or libraries. If we now come across a list of names and addresses, knowing that it was created from the identifying part of this register, we know (since we know about the register) that these persons are members of this minority, although neither the identifying part nor the "statistical part" of that register may contain any reference marker to that minority. This additional information is contained in the definition of the system. Let us further assume that suddenly we learn from other sources that members of this minority have a very high occurrence of a particular disease or a genetic defect. Then this address list is turning into a medical register that might be used as a list for potential quarantine measures. Even if we are not given names and addresses but only the statistical part of the register, we can plan such quarantine measures (or more clearly discriminating measures) with sufficient accuracy, since we know the geographical area (as part of the definition of our system) and the exact number, ages, and sexes of the persons to be evacuated from this area.[9]

The systems view

The attraction of PET concepts, and perhaps also one of their main purposes, is, as I have stated before, that they take the system designers' view of the world and talk to the designers in their own language. The system designers' view, however, is one of abstraction followed by formalization. This necessary "professional deformation" creates the risk of overlooking the *interconnections* of systems. Having solved the task of transforming a personal information system (or several of them) into a PET system may render these achievements doubtful if an undetected (or a deliberately untouched) non-PET system remains. Even more so, several PET systems may turn into less privacy-enhancing or into non-privacy-enhancing systems if they are linked in a certain way.

Turning a school planning system and a library planning system for a minority into a PET system would lose much of its privacy-enhancing effect if a subscription file for a paper that is exclusively read by the minority were to remain a non-PET system.

In sum: One would have to ensure not only that individual PET systems were designed properly but also that their role in networks of PET and non-PET systems remained clearly identifiable.

The technical assumption

Some PET systems are based on a specific technical assessment: the significant imbalance between the efforts to encrypt and the efforts to decrypt.

It is not clear to me to what extent this imbalance will hold in a mid-range time perspective.[10] Assessments of past technologies and subsequent developments seem to invite some caution.

External limitations

Resistant economic, social, and political forces have to be reckoned with when trying to implement PET concepts. Among the issues to be considered here are the economics of information, the needs of mobilization, and the concept of privacy itself.

The economics of information

Referring to the economics of information is restating the obvious. Personal information given away in a transaction process is part of the payment for the desired service or good. If this information is no longer there, the price of the good or service is likely to change. This change could be made transparent, and a "choice" could be offered: a company providing you with a good or service could ask a higher price that would include a higher degree of privacy of the information you provided, or you could pay a lower price but give additional information that might make the marketing strategies of that company more effective or that would allow that company to obtain extra revenue from that information. If competition does not allow for such changes, then perhaps habits of thinking will have to change in the advertising and direct-marketing industries. They will have to reconsider their assumptions about the use, the need, and perhaps the effect of personal information in the context of their activities. The difficulties of changing these habits should not be underestimated. Contrary to popular belief, advertising and direct marketing seem to be among the most conservative industries if judged by the amount of money these industries have spent so far on avoiding regulatory change.

Mobilization

These traditional economic approaches build on a more general social trait. In the process of social modernization, disturbances in social bonds (family, neighborhood, workplace, peer groups) are being compensated by strategies of mobilization in both the public and the private sector.

Governments, administrations, political parties, churches, companies, your grocery store, your neighborhood pizzeria, and your barber constantly call upon you *as an individual* to involve you in a relationship, since such relationships have become unstable and have to be renewed at every possible instant.[11]

The force of mobilization as a social need should not be underestimated. Mobilization answers not only the demand to involve but also the demand to be involved. Not only do governments seek to involve their citizens; citizens want to get involved (or at least feel involved) with their governments. Depending on how much bonding mobilization can offer, and depending on how strong the need for such bonds may be, anonymity as provided by PETs may well lose its attractiveness. On the other hand, privacy values may prove resistant to mobilization attempts, particularly if they are too blatant. This will most likely lead to two parallel developments: mobilization techniques will become more refined to overcome resistance generated from a desire for anonymity, and PET designs may have to contain doors or switches by which the subject may remain "reachable" provided certain conditions set by him or her are met.

The concept of privacy
The very concept of privacy on which PET design still largely seems to be built—privacy as anonymity, or in more advanced settings privacy as a conscious choice between anonymity and identification, or in even more refined systems privacy as an opportunity to freely variegate grades of anonymity—has its own limitations. Such an understanding of privacy misses a broader concept of privacy, a concept I want to call "political privacy." Political privacy regards the choices of anonymity as an integral *part* of a set of liberty rights, as one specific communication mode among others, and seeks to combine traditional privacy with more active participation-oriented elements.[12] Applying this more political understanding of privacy to PET concepts would have two consequences: First, PET design itself must be opened to participatory elements. This implies that designing PETs and implementing them in social systems must involve those whom these enhancements are supposed to serve. Second, PET designs would have to contain a broader variety of switches or modules that would allow for easily manageable situative choices. This flexibility

would not only serve the pressures of mobilization as described above; it would also integrate PETs into broader variety of political communication possibilities. Current political processes already contain a variety of such communication modes: for example, democracy as a social system of participation combines the possibilities of anonymous (secret ballot, anonymous political pamphlet) and identifiable (petition, taking the floor, private member bills, open voting) interventions. Any technical devices introduced into such processes should at least maintain such choices and make them easier to use and to change.

This is not to say that there is no place for or only limited use for more traditional PETs within the concept of political privacy. Rather, I would regard them as supplementary.

Privacy and the Role of PETs

Going back to one of the older texts in the field, a text published at about the same time that the first data-protection law (the Data Protection Act of the State of Hesse) took effect, I came across an article that pointed to the need to accompany such legislation not only with data-security measures but also with privacy-ensuring technical designs.[13] From what I remember, the German discussion that followed this aspect did not receive much attention. Even the basic rule that the best data protection is to have no personal data at all was mentioned only occasionally, almost resignedly, in the annual reports of data-protection agencies. In the years that followed, most of the attention went to spreading the knowledge of data-protection legislation internationally, and eventually, with some success and, with a lot of patience, to getting such legislation implemented. All efforts then focused on administering this legislation properly and on keeping up with technological developments.

Routine has moved in. In Europe, this routine threatens to keep us from re-evaluating our understanding of privacy. We have to look constantly at the way privacy is perceived in society (or, rather, in our societies), and we have to be prepared to reconsider our normative position. This is what juridical interpretation usually is about, but obviously this is not only a task of the juridical system. This is an obligation we have to accept whenever we set out to design technical systems in social environments. Against this background, it seems useful to have an occasional look at some of the other sensory mechanisms that allow us to monitor the understanding of privacy. After a long time during which this concept seemed to have received little attention in social sciences, in contemporary philosophy and in the arts this now seems to be slowly changing.[14] I want to point to three trends which I believe will be relevant to the future development of the concept of privacy and of PET or PET-like design approaches. The elements I wish to address in this context are information balance, identity, and trust.

Information Balance

With the notion of information balance I am not referring to the balancing of information *interests*. Balancing information interests is the more traditional concept that builds on the assumption of a continuous social contractual relationship in which balances have to be made between the interests of the "data subject" and the interests of other individuals (e.g. the data controller) and with societal interests (public security, the economic well-being of the state, etc.). Rather, by "balance" I am referring to adequately distributed information resources, to the accessibility of communication technology, and to the accessibility of the design process.

This is currently being discussed with regard to the telecommunications infrastructure, with regard to public-sector information resources, and with regard to mass media. We deal with issues of social cohesion, accessibility of information, and participation in electronic decision-making processes; we talk about the issues of electronic democracy. The relevance of the future development of PETs in this context seems to me to be twofold.

First, the design process of designing PETs has to include more directly individuals and their chosen forms of representation. I have already referred to the need to open up the traditional understanding of privacy to a political understanding of this concept. We might conceive PETs in this context as public utilities that invite interrogation, participation, and decision making processes that have been established for the control of the traditional public utilities. Data-protection agencies could play an important role in adopting such procedures for the design and implementation of large personal-information systems and their PET components.

Second, PET design should also direct its attention to participatory processes in the context of "electronic democracy." In the comments above on mobilization and the concept of privacy, I suggested that too narrow an interpretation tends to overlook the need for social involvement, for encouragement to become part of the political process, for contributing to social change by showing openly and identifiably dissent or non-integrated behavior—in short, that it overlooks the need for the display of civil courage. On the other hand, in a recent decision the US Supreme Court has restated not only the legitimacy but also the political importance of "anonymous voices" in the political debate.[15] What may be needed, then, is a more flexible management of the "digital persona"[16] for the participating political individual.

Identity

One of the more recent critiques of the notion of privacy argues that it misses the essence of identity. Identity is not a constant but a process. We do not develop our identity by keeping ourselves separate from others; our identity is what others know about us. In knowing about us, power is already exercised. The way we are seen by others greatly influences the way we see ourselves. Information in that sense means forming our identity. In order to ensure emancipation under such conditions, we must already intervene in the process of subject formation through information.[17] As familiar as this concept may sound when looking less at recent social theory than at less recent literary contributions,[18] it nevertheless opens a perspective on a variety of modifications for PETs, again aiming at giving us a broader variety of establishing and displaying different aspects of ourselves. This leads us back, in this case outside the political process, to PETs as systems for managing our various displays of personality in different social settings. PETs might turn into electronic managers of identity agents, keeping track of what we reveal about ourselves to whom and under which conditions. PETs would then help us to remember whether, when, and toward whom, on the Internet, we had been a dog or a cat.

Trust

One of the main arguments against the broader implementation of PETs is the need or the habit of meeting the insecurity of life with greater confidence and of maintaining or regaining certainty in a world of uncertainty. I will not go into an analysis as to whether this is an inherent, almost natural trend in the rationalization and industrialization processes of our societies, creating so many risk situations that constantly seem to breed new security demands.

One possible strategy to respond to this development seems to be to plead for looser relationships between social entities, to make them less dependent on one another, and to regard criteria like efficiency which place all social systems under the same logic from a more relative perspective.[19] In such a model of a society, trust would have to play an essential role.[20]

By trust I am not referring to trust as in "trusted third party" or in "trusted systems." In these contexts, "reliability" seem to be a more appropriate term: these systems answer certain requirements within a given probability; they are deemed to be reliable. Such systems, however, then need no longer to be trusted; they are safe within that given probability; they involve no or at least only calculable risks. Trust as a social phenomenon, however, involves a conscious decision to interact *although* there is risk. Relationships in which risk taking is eliminated or essentially reduced by devices, whether of a technical or a social nature, leave little room to display trust. They also provide few opportunities to be trusted and thus to feel appreciated, needed, and involved. The less these feelings can occur the looser the bonds that hold a society together may become.

Trust in this sense will have to play a larger role in the design of social systems and information systems in particular. This is not only a normative statement. It is also a prediction. Moving away for a moment from the state of paranoia that the word "surveillance" often invokes,[21] I want, for a moment, to invite the reader to an experiment in metanoia[22] to see the point: If one looks at breaches of computer security from a different perspective, eliminating cases of negligence, malicious intent, and sheer stupidity (trust has to be consciously selective), one is left with cases that in fact have been displays of trust. One tends to look at such cases with ridicule if not with contempt. It may well be that we will have to change our attitude toward such displays of trust and concentrate our contempt and ridicule more on those who failed the trust than on those who trusted. From this perspective, such cases make us aware how frequent the display of trust still is in electronic environments; at least it is far more frequent than the discussions of the surveillance society seem to indicate.

Trust is also, perhaps, the answer to the well-known surveillance paradox. (Surveillance must be controlled; however, each layer of surveillance calls for another, and this leads to an infinite regression: surveillance of surveillance of surveillance, and so on.) The capability of people to insert into these systems acts of conscious risk taking in social relationships (i.e., trust) ensures the functioning of society in spite of these mechanisms. PETs, if designed in a manner that allows us with ease to consciously and selectively reveal our identities may help us to give social mechanisms of establishing and maintaining trust a broader space. They might encourage us to at least consider interacting more often in uncertain environments.

Conclusion

We might regard PETs as a technical innovation to help us to solve a set of socio-political problems. Their implementation—and this may turn out to be their most important feature—forces us to return to *social innovation* in order to successfully implement them. This then leads us to our main task as social scientists, lawyers, regulators, and privacy practitioners: to accept the challenge of information and communication technology as a challenge for social innovation. Data-protection regulation has already been a historic case in social innovation. With the help of PETs, there are further challenges to be met.

This chapter is a revised version of a presentation made at the conference "Visions for Privacy in the 21st Century: A Search for Solutions" (Victoria, British Columbia, 1996).

Notes

1. For further details, see the laudable cooperative achievement of the Netherlands and the Ontario data-protection agencies: Information and Privacy Commissioner (Ontario, Canada) and Registratiekamer (Netherlands), *Privacy*

Enhancing Technologies: The Path to Anonymity, volumes I and II, 1995. For the cryptographic issues see also Bruce Schneier, *Applied Cryptography*, second edition (New York, 1995).

2. For an example of such a design see Dag Elgesem, Privacy, respect for persons and risk, in *Philosophical Perspectives on Computer-Mediated Communication*, ed. C. Ess (New York, 1996).

3. Spiros Simitis, Lob der Unvollständigkeit—Zur Dialektik der Transparenz personenbezogener Informationen, in *Gegenrede: Aufklärung–Kritik–Öffentlichkeit*, ed. H. Däubler-Gmelin et al. (Baden-Baden, 1994), pp. 573–592.

4. See *Privacy Enhancing Technologies*, volume II, p. 26 ff.

5. This is clarified on page 27 of volume II of *Privacy Enhancing Technologies*.

6. *Official Journal*, series L, 281/31, November 23, 1995.

7. Article 2(a) of directive.

8. See Herbert Burkert, Die Eingrenzung des Zusatzwissens als Rettung der Anonymisierung? *Datenverarbeitung im Recht* 8 (1979): 63–73.

9. Although to some readers this example may sound somewhat familiar, it is still purely hypothetical.

10. See Steven Levy, Wisecrackers, *Wired* 4.03 (1996): 128.

11. Regis McKenna, *Relationship Marketing* (London, 1991).

12. Bas van Stokkom (Citizenship and privacy: A domain of tension, in *Privacy Disputed*, ed. P. Ippel et al. (The Hague, 1995), p. 53 ff.) makes an anecdotal reference to Louis Brandeis's intending to publish "The duty of publicity" after "The right to privacy."

13. Wilhelm Steinmüller et al., *EDV und Recht. Einführung in die Rechtsinformatik* (Berlin, 1970), p. 88.

14. For an overview on trends in political science and contemporary philosophy see Colin Bennett, The political economy of privacy (1996). For examples in the arts, see e.g. the works of Julia Scher.

15. *MacIntyre v. Ohio Elections Commission*, 115 S.Ct. 1511 (1995); *Figari v. New York Telephone Company* 303 N.Y.S.2d. 245 (App. Div. 1969). See also Henry H. Peritt, *Law and the Information Superhighway* (New York, 1996), section 6.2.

16. Roger Clarke, The digital persona and its application to data surveillance, *Information Society* 10 (1994), no. 2.

17. See Marc Poster, *The Second Media Age* (Cambridge, 1995) and *The Mode of Information* (Chicago, 1990).

18. In particular, the works of Fernando Pessoa (1888–1935).

19. Colin Bennett, The Political Economy of Privacy: A Review of the Literature, unpublished paper, April 1995, p. 30 ff.

20. Herbert Burkert and Murray Rankin, The Future of the OECD Privacy Protection Guidelines: Building Trust in Electronic Data Networks (ICCP), Paris, OECD, June 26, 1989; Herbert Burkert, Electronic Trust and the Role of Law: A European Perspective, in *13th World Computer Congress 1994*, ed. K. Brunnstein, volume 2.

21. See David Lyon, *The Electronic Eye: The Rise of Surveillance Society* (Minneapolis, 1994), p. 218 ff.

22. Peter M. Senge, *The Fifth Discipline* (New York, 1990).

5

Re-Engineering the Right to Privacy: How Privacy Has Been Transformed from a Right to a Commodity

Simon G. Davies

The concept of privacy has shifted in the space of a generation from a civil- and political-rights issue motivated by polemic ideology to a consumer-rights issue underpinned by the principles of data protection and by the law of trading standards. Privacy has metamorphosed from an issue of societal power relationships to one of strictly defined legal rights. Several mechanisms have played important roles in this shift: Opposing players are being recast as "partners" in surveillance. Private right and public interest have been subtly but substantially redefined. Privacy invasion is often accompanied by the illusion of voluntariness. The number of public campaigns in response to a growing number of surveillance initiatives by private- and public-sector organizations has plummeted as traditional privacy activism has declined. This chapter discusses the changing nature of privacy and the new rules that define public interest and the private right.

If opinion polls accurately reflect community attitudes, concern over privacy violation is now greater than at any other time in recent history. Populations throughout the developed world report anxiety about loss of privacy and about the surveillance potential of computers. Despite the perception expressed in opinion polls that surveillance is flourishing, privacy activism is losing momentum. In many countries—particularly throughout Europe— the privacy movement has been inactive for almost a decade. Privacy advocacy has been recast as a legal and a consumer-rights issue.

The history of privacy activism is by no means uniform, but in many countries it is clear that the concept of privacy has in previous decades been interpreted as a political idea fueled by issues of sovereignty, technophobia, power, and autonomy. In the present-day context—in those same countries—privacy protection is widely perceived as constituting a set of technical rules governing the handling of data. While there are consequently more codes, conventions, and laws in place than ever before, more data on more people is being collected by more powerful systems and for more purposes than at any other time in history. Despite the evolution of technologies that were the subjects of nightmarish dystopias in the 1960s, resistance to their introduction has been minimal. There are some limitations on information systems, but all developed countries now have the technical capacity to numerically code and geographically track large numbers of their people. Government organizations have the ability to collect, store, cross-correlate, and cross-match intimate details about key aspects of lifestyle, activities, and finances.

The Big Brother society that was imagined in 1970 depended on coercion and fear. The society we are developing now appears to be more Huxley-like than Orwellian. It is a Brave New World dominated not so much by tyranny as by a deadening political and cultural phenomenon that Ralph Nader calls "harmony ideology."[1]

What appears to have occurred since 1980 is a shift in the perception of privacy and privacy invasion, rather than a diminution of public concern. The effect is an ambivalence that retards consumer and political activism over even the most blatant privacy intrusions. In many countries, fundamental changes have taken place in society's approach to traditional privacy issues. Five factors can be identified as central to this change:

from privacy protection to data protection   Formal rules in the form of data-protection principles appear to have satisfied some of the concerns of information users and the public, but have failed to stem the growth of surveillance.

the creation of partnerships   All stakeholders, whether proponents of surveillance or traditional opponents, have been transmogrified into a "partnership" with common goals and desires.

the illusion of voluntariness   Many surveillance schemes now involve a "voluntary" component that has the effect of neutralizing public concern.

privacy rights as commodities   Many traditional rights have been put on a commercial footing, thus converting privacy rights into consumer issues (as in the case of Caller ID blocking) *the concept of "public interest"*   This concept appears to have expanded substantially in recent years, to the point where in some countries it is now a generic description for all activities of the state.

These trends are fundamentally changing the nature, the scope, and the relevance of privacy.

History

In the 1960s and the early 1970s, as information technology emerged as a tool of public administration, large computer projects encountered a groundswell of fear and resistance. In Europe, those who had experienced Nazi occupation were not inclined to trust the idea of centralized databases.[2] In the United States, the citizenry of which enjoys a long tradition of distrust of government, information-technology planners in Washington had to tread with caution. A proposal for a US National Data Center had already been debated and abandoned some years before. Proposals to create a national identification card had suffered the same fate.

In Britain, concern about the establishment of government databases was at a similar level as concern over unemployment and education.[3] In several European countries, legislation and constitutional provisions were passed which set conditions on the use of information, and which even prohibited schemes such as national "population numbering."[4] A substantial number of citizens in all industrialized countries looked forward with trepidation to the impending Orwellian dystopia. Alan Westin described the prevailing mood as "profound distress."[5]

Attitudes changed between 1965 and 1990. Since the mid 1980s, opinion polls reveal a high level of concern about computers, but this is rarely translated into political action. Most of the population has nurtured a symbiosis with information technology. Many consumers have been prepared to surrender their personal information to information systems in return for the promise of a safer, cheaper, more efficient life.

Almost all countries have seen an increasing range of surveillance activities undertaken by government and private bodies. In Britain, Parliament has passed the Criminal Justice Act, mandating the establishment of a national DNA database, limitations of the rights of freedom of movement and assembly, a requirement to prove identity in numerous circumstances, and a significant expansion of police powers. A national identification card is almost certain to be announced before the 1997 election, and data-matching programs are already underway among government agencies. In January 1996, the Department of Social Security proposed a national registry of "palm prints" for all recipients of government benefits without negative reaction from the press or the public. The governing and the opposition party both support a general curfew for young people. Meanwhile, Caller ID and extensive electronic visual surveillance (via closed-circuit television) have been introduced with minimal comment from civil liberties and consumer groups.

The situation is replicated in the United States, which currently faces a wide range of surveillance initiatives, including a national employment-verification system, invasive health-data systems, a national DNA database, Internet censorship, and controls on the use of encryption. Protection from such activities is patchy at best. Since the 1970s no omnibus privacy initiative has mustered enough support to even get out of the committee stage, but interest in privacy advocacy in Congress may currently be lower than at any other point in recent history.[6]

In Australia, an integrated national tax number and an unparalleled regime of data matching among government agencies has created a single national data system. It is difficult to reconcile this state of affairs with the fact that in 1987 millions of Australians participated in a protest against a proposed national ID card.

In 1986, the federal government announced its intention to introduce the "Australia Card." This credit-card-size document was to be carried by all Australian citizens and permanent residents. (Cards in a different format would be issued to temporary residents and visitors.) It would contain the person's photograph, name, unique number, and signature and the card's period of validity. The original purpose of the card was to combat tax evasion, but so popular was the idea that the number of government agencies wanting to use the card leaped in 6 months from 3 to 30. The purpose of the proposed card was extended to include catching assorted criminals and illegal immigrants and to being used for financial transactions, property dealings, and employment. Health benefits, passport control, and housing were then added. At first the public accepted the idea of an ID card without question, but in mid 1987 a spontaneous campaign of opposition was initiated under the banner of a diverse coalition called the Australian Privacy Foundation.

By August the government was facing an internal crisis over the card. The left of the Labor Party had broken ranks to oppose the card, while right-wing members (particularly those in marginal seats) were expressing concern within their caucus. In September, in the face of massive public protests and a growing split in the government, the plan was scrapped. A that moment, privacy was at the top of the political agenda. Australians began to reinterpret their culture through a framework of privacy.

Although it is true that consumers have developed a close relationship with computers, it is equally true (even today) that they are worried by surveillance. In the United States, national polls reveal increasing concern over privacy. A 1995 poll conducted by Louis Harris and Associates ranked consumers' concerns about privacy rated slightly behind their concerns about controlling false advertising and reducing insurance fraud; 82 percent of Americans were found to be "very concerned" or "somewhat concerned" about privacy.[7] These results paralleled those of a Yankelovich poll which found that 90 percent of Americans favored legislation to protect their privacy from invasion by businesses.[8] The Harris poll has been conducted regularly since 1976, at which time the "very concerned" or "somewhat concerned" rate was 47 percent. In 1983 it was 76 percent.[9]

In Australia, a recent Morgan Gallup poll ranked privacy second in importance only to education. And yet seven out of ten respondents expressed the view that privacy no longer exists because (they believe) the government can find out whatever it wants about them.[10] The 1995 Harris poll also found that 80 percent of Americans believe they have lost all control over personal information.

The citizens of Western industrialized countries want privacy, but feel it is extinct. They are aware of the loss of privacy, but feel powerless to defend themselves against intrusive practices. These feelings may be due in part to the increasing difficulty of defining privacy rights. A broad range of other interests—many of them "public" interests—compete with privacy.[11]

The Great Angst: Personal Concerns That Compete with Privacy

Concerns about computers, surveillance, and privacy invasion may be widespread, but they account for only a small part of the anxiety spectrum. Opinion polls reveal that anxiety level has not only intensified but also broadened since the 1970s. Anxiety is being experienced across a wider range of issues, some of which are new. Privacy is dwarfed (and sometimes neutralized) by these newer concerns the foremost of which are employment and crime.

In many developed nations, particularly in Europe, anxiety over employment has become endemic. Most of the population is either unemployed or engaged in employment which is unstable or short term. The British commentator Will Hutton has labeled this the "30-40-30 society": 30 percent are unemployed or work part time, 40 percent have tenured jobs, and 30 percent are in unstable, short-term employment.[12] Explaining the situation in the United States, Lance Morrow observes that "America has entered the age of the contingent or temporary worker, of the consultant and the subcontractor, of the just-in-time work force— fluid, flexible, disposable."[13] It is a telling statistic that Manpower, with a "staff" of 560,000, is now America's largest employer.[14] *The Economist* notes: "People who used to think their jobs were safe and their standard of living would rise each year can no longer take either for granted. . . . In many economies, competition (domestic as much as foreign) and new technology are touching people who were hitherto immune from such forces."[15]

Anxiety is high among employees of all types. In the United States, three-fourths of the part-timers who work all year receive no health insurance, and fewer than 20 percent are included in a pension scheme.[16] In the general workforce, 75 percent of employees describe their jobs as stressful.[17]

These anxieties have been caused by factors too numerous to discuss here, but among the most important is the current fashion for "re-engineering" companies. Consultants and accountants have radically altered many organizations culturally and structurally. They have reorganized, de-layered, abolished departments in favor of profit centers, instituted short-term employment contracts, set up internal markets, and created a mass market for outsourcing. The result is a redefinition of employment that many workers find hard to accept.

Anxiety is created not only by individual threats but also by rapid change and evolution. The Australian social researcher Hugh Mackay concludes: "The Australian way of life has been challenged and redefined to such an extent that growing numbers of Australians feel as if their personal identities are under threat as well. 'Who are we?' soon leads to the question 'Who am I?'"[18]

A second fundamental anxiety has continued to grow in the past 20 years: that of personal security. Concern over the threat of crime has always been a feature of urban society, but rarely has it been so consistently accompanied by anxiety over the integrity of law-enforcement and judicial institutions. Such concern is often out of any reasonable proportion to the actual extent of crime.[19] In many countries there is a parallel perception that the police and the courts have become part of the crime problem. In these circumstances, invasion of privacy is seen as a price worth paying for the promise of security. Privacy becomes a matter of "balance" rather than a fundamental human right. Of course, it could be argued that most fundamental rights are subject to the same balancing act. Freedom of expression, freedom of assembly, and freedom of movement are subject to conditions of public interest.

Public insecurity is fed by lurid and uncritical reporting in the news media, which leads in turn to reactive and heavy-handed security measures. The result is a widening array of security procedures directed against ever-broadening definitions of who and what constitutes a potential threat.[20] The Los Angeles Forum for Architecture and Urban Design observes: "This perceived insecurity has inspired urban paranoia manifested as belligerent territorialistic reactions on the part of more privileged groups against less privileged and more marginalized sectors of society. Publicly accessible spaces are made progressively harsher. They are stripped of amenities, and lined with blank walls, bathed in high intensity security lighting, and studded with observation cameras and ubiquitous placards warning of 'armed response.'"[21] Privacy has little place here, either as a right or as a value. It is manifested as solitude, and as such it must be a pre-designed component of urban design.

Privacy Paralysis in Practice: Closed-Circuit Television in Britain

The new generation of closed-circuit television (CCTV) surveillance equipment comes closer to the traditional perception of Big Brother than any other technology. There are now linked systems of cameras with full pan, tilt, zoom, and infrared capacities. Among the sophisticated technological features of these systems are night vision, computer-assisted operation, and motion detectors that place the system on red alert when anything moves in view of the cameras. The camera systems increasingly employ bullet-proof casings and automated self-defense mechanisms. They can be legitimately described as military-style systems transplanted to an urban environment.[22] The clarity of the pictures is often excellent—many systems are able to recognize a cigarette pack at 100 meters. The systems can often work in the dark, bringing images up to daylight level. The most sophisticated systems can store and process images in digital form. Experiments in automated computerized face recognition have reached an advanced stage.[23] If any technology was to provoke the ire of a community, this should be the one. In Britain, however, the reverse is true.

The use of CCTV in the UK has grown to unprecedented levels. Between 150 million and 300 million pounds (between 230 and 450 million dollars) per year is now spent on the surveillance industry, which is estimated to have 200,000 cameras covering public spaces.[24] According to the British Security Industry Association, more than 75 percent of these systems have been professionally installed. Almost 200 towns and cities have established sophisticated CCTV surveillance of public areas, housing estates, car parks and public facilities.[25] Growth in the market is estimated at 20 to 30 percent annually. In his speech to the October 1995 Conservative Party conference, Prime Minister John Major announced that the government would fund an additional 10,000 cameras. A Home Office spokesman recently told a seminar at the London School of Economics "if this all saves just one life, it's worth it."[26]

Many central business districts in Britain are now covered by surveillance camera systems. Their use on private property is also becoming popular. Increasingly, police and local councils are placing camera systems into housing estates and red light districts.[27] Residents' associations are independently organizing their own surveillance initiatives. Tens of thousands of cameras operate in public places—in phone booths, vending machines, buses, trains, and taxis, alongside motorways and inside bank cash machines.

The government has placed video surveillance at the center of its law-and-order policy. This form of surveillance is likely to extend even to the home. Andrew May, Assistant Chief Constable of South Wales, has urged victims of domestic violence to conceal video cameras in their homes to collect evidence.[28] The technology is already being used in hospitals to support covert surveillance of parents suspected of abusing their children.

The limits of CCTV are constantly being extended. Originally installed to deter burglary, assault, and car theft, most camera systems have been used to combat "anti-social behavior," including such minor offenses as littering, urinating in public, traffic violations, fighting, obstruction, drunkenness, and evading meters in town parking lots. They have also been widely used to intervene in underage smoking and a variety of public-order transgressions.[29] Other innovative uses are constantly being discovered. The cameras are particularly effective in detecting people using marijuana and other substances. According to a Home Office promotional booklet entitled CCTV: Looking Out for You, the technology can be a solution for such problems as vandalism, drug use, drunkenness, racial harassment, sexual harassment, loitering, and disorderly behavior.[30]

CCTV is quickly becoming an integral part of crime-control policy, social control theory, and "community consciousness." It is widely viewed as a primary solution for urban dysfunction. It is no exaggeration to conclude that the technology has had more of an impact on the evolution of law enforcement policy than just about any other technology initiative in the past two decades.

In a domain of public policy which the public generally believes is short on solutions, CCTV is a highly attractive investment. When fear of crime and criminals is intense, proponents of surveillance routinely portray critics of surveillance as enemies of the public interest.[31] The effectiveness of CCTV in preventing crimes is not certain, but it would be difficult to deny that the technology is quickly changing the face of crime prevention and social control. With these changes come important shifts in relationships between people and institutions.

Proponents of CCTV are inclined to describe opposition to visual surveillance as marginal. Although it is certainly true that the majority of the UK's citizens support CCTV as a crime-control measure, this support is by no means unconditional. In one survey commissioned by the Home Office, a large proportion of respondents expressed concern about several aspects of visual surveillance. More than 50 percent of respondents felt that neither the government nor private security firms should be responsible for the installation of CCTV in public places, 72 percent agreed that "these cameras could easily be abused and used by the wrong people," 39 percent felt that the people in control of these systems could not be "completely trusted to use them only for the public good," and 37 percent felt that "in the future, cameras will be used by the government to control people." Although this response could be interpreted a number of ways, it goes to the heart of the dilemma of privacy versus civil rights. More than 10 percent of the respondents believed that CCTV cameras should be banned.[32] Contrast this with the view of Chief Constable Leslie Sharp of Scotland's Strathclyde Police Department. Referring to his force's strong commitment to CCTV, he told ABC's "20/20" program: "We will gradually drive the criminal further and further away, and eventually I hope to drive them into the sea."[33] Few residents of his region openly oppose this rationale.[34]

All five of the core elements of change are evident in the above scenario. The government has taken a lowest-common-denominator response to regulating—but not controlling—the cameras. The Code of Practice that has been published sets out operating standards but has no mechanism for accountability or enforcement.[35]

Defining the Private Right, the Public Interest, and the Question of "Balance"

Britain's epidemic of CCTV highlights the unruly and contradictory nature of privacy protection. In matters of "public interest," privacy tends to be whatever is left after more pressing elements have been resolved. Whether through design or osmosis, information users employ a common set of terms that are hostile to privacy. In the parlance of banks, police, and government agencies, privacy is a value rather than a right. It is a commodity that imperils efficiency and shields criminals. It is the natural enemy of freedom of information. It is a concept without definition or form. Its enforcement is cumbersome and expensive. Thus, privacy is cast as the *bête noir* of law enforcement, openness, progress, efficiency, and good government.[36]

Privacy invasion and its related technologies are promoted in neutral language. In Australia, the national ID card became the Australia Card. The Cash Transactions Reporting Agency, responsible for tracking the use of cash, had its name changed to the less provocative AUSTRAC. The imaginatively named vehicle-tracking system Scam Scan was quickly changed to Safe-T-Cam to avoid any implication that it might in the future be used for something other than the detection of speeding truck drivers. And, after a controversy over "smart cards" in the health sector, the cards became Chip Cards and, later, Service Cards and Relationship Cards. Customer-profiling systems were marketed as Loyalty Clubs.[37]

This language is predicated on a set of widely held beliefs. Among the most important of these are "technology cannot be stopped," "only those with something to hide have something to fear," and "all science is neutral." In some quarters, these have become axiomatic truths.

Another weapon in the armory of information users is the multiplicity of definitions in use. The pursuit of a single definition of privacy has preoccupied so many travelers in this field that the quest has become a standing challenge in the privacy field. Even after decades of academic interest in the subject, the world's leading experts have been unable to agree. One pioneer in the field described privacy as "part philosophy, some semantics, and much pure passion."[38] There are many reasons for this state of affairs. Privacy is a generic label for a grab bag of values and rights. To arrive at a general definition of privacy would be no easier today than finding consensus on a definition of freedom. The subject must be brought into sharper focus.

The Quantum Shift in Privacy Activism

For most of the 1960s and the 1970s. fueled by a strong spirit for protection of democratic rights, academics felt comfortable exploring views that in future years (in most circumstances) would be viewed as extreme.

In the early 1970s, with the advent of new information-gathering techniques, a strong anti-census movement evolved throughout Europe. The protest in the Netherlands was so widespread that it achieved a critical mass that finally made the census entirely unworkable. A substantial number of Dutch citizens simply refused to supply information to the census authority.[39] At that point in Dutch history, privacy achieved a notable place on the political agenda. In the wake of the census protest, a new organization called Stichting Waakzaamheid Persoonsregistratiie (meaning Privacy Alert) was formed. For the next 20 years it provided a powerful and effective focus for privacy issues in the Netherlands.

By the late 1980s, Stichting Waakzaamheid Persoonsregistratiie had become the world's best-staffed non-government privacy watchdog. However, in 1993 its fortunes went into decline, and in 1994 the organization was dissolved. Although the decision of the Board of Directors to close down the organization involved financial matters, it also reflected changes in social attitudes toward privacy. It seems that the board was simply not committed enough to support the group through hard times, despite clear evidence that Holland was facing severe and widespread privacy problems. Indeed, the end of Stichting Waakzaamheid Persoonsregistratiie came at a time when the country was establishing a single national numbering system, an identity card, and a vast array of new police and administrative powers motivated by the Netherlands' partnership in the Schengen Agreement[40]—issues far more fundamental than the census. According to the Chip Card Platform, which is coordinating the ID card, the political change in recent years highlights "an understanding among the public of the role of information technology." Officials do, however, acknowledge that this change of attitude has taken them by surprise.[41]

The Dutch group was easily the world's most successful non-government privacy organization, yet it failed to survive the most crucial period of its history. Other organizations elsewhere have suffered the same fate, but alternative approaches to privacy protection have evolved in their place. The Canadian Privacy Council, despite a promising inauguration in 1991, has failed to materialize. However, in 1996 the Canadian Standards Association released a groundbreaking formal privacy standard.[42] The New Zealand Privacy Foundation, born in the heat of the campaign against the Kiwi Card, now exists only in name,[43] although the existence of the foundation motivated the government to enact a relatively strong and broad privacy law. The Australian Privacy Foundation, which organized the massive 1987 campaign against the ID card, now serves only as a response mechanism, with few members and no budget. In its place is the Australian Privacy Charter, which—not unlike the Canadian situation —provides a more quantifiable approach to the privacy issue.[44]

The loss of Stichting Waakzaamheid Persoonsregistratiie is an important symbol of changing times. With the sole exception of the Washington-based Electronic Privacy Information Center (EPIC), privacy lobbies around the world are becoming less effective. Shifts in public attitude have created new and complex challenges that privacy groups have yet to absorb.[45] The transmogrification of privacy rights into legal and consumer rights means that the slack is being taken up by institutional bodies, such as courts, industry watchdogs, and trading-standards bodies. That is not to say that many individuals and community groups are not responding to privacy issues. The Internet and related technologies offer fertile ground for new forms of privacy activism. Among these are EPIC's electronic petition against the US government's Clipper Chip proposal,[46] the use of web sites as a means of exposing the private lives of public figures who have opposed privacy protection,[47] and the use of electronic mail in "wildcat" strikes against privacy-invading organizations.[48] Privacy International and other groups have used the Internet to reach a broader audience of experts and supporters.

Some recently published critiques of technology and information,[49] together with a broadening range of popular and analytical works on privacy, have contributed to a reopening of the privacy discussion. The result is that an increasing proportion of the public is becoming aware of complex privacy issues.

Assessing the Five Fundamental Transitions in Privacy

It is tempting to assume that the demise of privacy is merely a sign of a natural shift in values. It is more likely that there are numerous causative factors which have been engaged by private and government interest groups, by legal and intergovernmental organizations, and by the media.

From Privacy Protection to Data Protection

As European governments and such international organizations as the OECD, the European Commission and the Council of Europe struggled from the early 1970s on to resolve growing fears about computers, a group of principles slowly evolved to form a basis for law. Data, according to these principles, should be collected, processed, and distributed fairly, accurately, on time, and with a measure of consent. Most were developed in consultation with—and with the approval of—information users in the private and public sectors.

In theory, the privacy of personal information is protected through these principles. In reality, such principles, and the regulators who enforce them, have a limited impact on key aspects of surveillance.

There is no question that some of the more blatantly intrusive activities of government departments are reined in as a result of the existence of data-protection laws. Individual citizens are given some protection over the way personal information is used. In theory at least, they can gain access to many of their files. However, even within this narrow scope, data-protection acts generally have serious limitations. One of the broadest deficiencies is that they are seldom privacy laws. They are information laws, protecting data before people. Instead of being concerned with the full range of privacy and surveillance issues, they deal only with the way personal data is collected, stored, used and accessed. In Britain and Australia, for example, these laws are generally not concerned with visual surveillance, drug testing, use of satellites, or denouncement campaigns (e.g., hotlines for reporting tax dodgers). Additionally, many data-protection acts do not cover publicly available information (that is, information and records, such as land titles and electoral rolls, that are available for general public inspection).

There are two particularly serious problems associated with the core principles. The first and most obvious is that they tend to allow a great many privacy violations to occur through exemptions for law enforcement and taxation. The second and perhaps the graver problem is that data-protection law does almost nothing to prevent or limit the collection of information. Many acts merely stipulate that information has to be collected by lawful means and for a purpose directly related to a function or activity of the collector. Thus, a virtually unlimited number of information systems can be established without any breach of law.

Despite some evidence that the adoption of these principles has calmed some of the fears of the public, it would be a mistake to assume that they will address the most pressing privacy problems. The Dutch privacy expert Jan Holvast recently explained that privacy legislation "corrects the mistakes and misuses but it does not attack the way in which technology is used. On the contrary, experiences with data-protection law in several countries show that these laws are legalizing existing practices instead of protecting privacy."[50] Privacy International observed in its 1991 report: "Protections in law, where they exist, are sometimes ineffective and even counter-productive. Extensive information holdings by government are invariably allowed under exemptions and protections in law. The existence of statutory privacy bodies, rather than impeding such trends, sometimes legitimates intrusive information practices."[51]

Even where privacy acts are governed by a dedicated and forceful regulator, the opposing forces are so overwhelming that it is often impossible to stand in the way of the most sophisticated data practices. David Flaherty, now Privacy Commissioner of British Columbia, says that "there is a real risk that data protection of today will be looked back on as a rather quaint, failed effort to cope with an overpowering technological tide."[52]

Without considerable positive discrimination in support of privacy interests, a privacy commissioner can do little. Flaherty has observed that "the public is being lulled into a false sense of security about the protection of their privacy by their official protectors, who often lack the will and energy to resist successfully the diverse initiatives of what Jan Freese (one of Europe's first data-protection commissioners) has aptly termed the 'information athletes' in our respective societies."[53]

Subjects of Surveillance Are Becoming Partners in Surveillance

Britain's CCTV schemes, together with those of other countries, have embraced an important promotional element. To give the systems added weight and appeal, they are invariably promoted as partnerships, with all stakeholders recast as investors or shareholders. These new partners usually include police, local businesses, the town council, community groups, media, the insurance industry, and "the citizens." This strategy parallels other schemes, such as "partners against crime" and "community partnership."

The flavor of most anti-crime advertising in Europe and America is one of togetherness: "Crime . . . together we can crack it" and "Working together for a crime free America" are typical. Partnerships are also a common element in "neighborhood watch" schemes, which require a certain level of participation.

The "partnership" or "shareholder" image is also becoming popular in other areas of privacy invasion. Health data networks uniformly adopt such a model. Australia's national Health Communications Network (HCN) envisioned a range of partners, including patients, investors, government, and researchers. The same model was used to promote Britain's NHS Wide Network and some US proposals for linked health-data networks.[54]

The designers of Australia's HCN project believed that privacy concerns constituted the key threat to the scheme's viability in the marketplace. The "privacy problem," as members of the project team described it, would have to be comprehensively addressed to maintain political and consumer confidence in the strategy. The "privacy problem" was dealt with in three ways. First, the project pursued the adoption of a set of strictly defined data-protection guidelines. Second, the images that represented the main areas of concern were redefined. For example, the concept of databases, which had always been at the heart of privacy fears in Australia, had to be recast in unthreatening terms. In a paper entitled "Your privacy or your health," one of the HCN's architects, Dr. Larry Cromwell, eliminated the database from the equation, arguing that in the HCN the database would be no more than a communication medium. The third device used to neutralize privacy was to turn every health consumer into a partner in the new system. (The most crucial stakeholders— the doctors—were the first to oppose the scheme.[55])

This reinterpretation of the stakeholder model brings together parties who traditionally would have been in opposition. In the "investor" or "partner" process, these parties are part of an inclusive formula that embraces all the major elements of a project. Inherent in the model is an implication that all stakeholders are integral to planning, are equal partners in the outcome, and are overall winners in the scheme of things. Of equal importance is an implication that contributions from non-stakeholders are invalid.

The Illusion of Voluntariness

Many surveillance schemes now involve a "voluntary" component which has the effect of neutralizing public concern about surveillance.

When the Cardiff police were hunting the murderer of a young girl in 1995, they asked the entire male population of a local housing estate to "volunteer" for DNA testing—"just so we can eliminate you from our enquiries," each was assured. The reality was that anyone who did not "volunteer" was considered a suspect and was therefore subject to special scrutiny.[56] In the same year, this tactic was used by the London police after the rape of a girl in Great Portland Street. In that case, police had written to local residents with certain physical characteristics, again arguing that volunteering for the test would "eliminate" them from further enquiries.[57] Also in 1995, the drivers of all Mercedes trucks of a certain color throughout England were subject to the same "request."

In some regions of surveillance, governments seem less inclined to make privacy invasion mandatory, choosing instead to say that participation is a matter of free choice in an open market of services. Shortly before the British government was to issue a Green Paper on a national ID card, cabinet papers revealing that the dominant view among planners was that police and civil-rights concerns could be resolved if the ID card was made "voluntary" were leaked.[58] In May 1995, the Home Office released its Green Paper on the ID card. The document offered numerous models for a card scheme, including voluntary cards, multi-purpose cards, and compulsory cards, in several formats. No particular format was recommended, though the document appeared to give special weight to a multi-purpose system compulsory only for benefit claimants and drivers (i.e., 90 percent of the population). For the remainder, the card would be "voluntary."

The British government appears to have taken a lead from the Australian experience with ID cards. In Australia, the architects of a national ID card used the expression "pseudo-voluntary." Although it was not technically compulsory for a person to obtain a card, it would have been extremely difficult to live in society without one.[59] There is some anecdotal evidence that this pseudo-voluntary approach may have the effect of neutralizing privacy concerns. It might be widely viewed that those who do not "volunteer" bring problems upon themselves.

This prospect could be exemplified by a international biometric hand-print-registration system for passport holders. If such a system were to be imposed by force, it would most likely result in a political scandal. Instead, it has been introduced with great success as a voluntary system. The project, called INSPASS (Immigration and Naturalization Service Passenger Accelerated Service System), has been operating since 1993 as a voluntary system for frequent travelers. More than 65,000 travelers have so far enrolled in the system, a figure that increases by almost 1000 a week. Governments in 26 countries are coordinating with the project.

If the INSPASS trial is successful, the technology may ultimately make conventional ID cards and passports redundant. In exchange for faster processing, passengers will have to accept a system that has the potential to generate a vast international traffic in their personal data. Ultimately, a universal immigration control system may be linked to a limitless spectrum of information, including the data in police and tax systems.[60]

It is ironic, in view of even a notional element of privacy, that people tend not to support more privacy-friendly technological options that are less likely to collect potentially damaging personal data. According to several sources, the signs are not good. Customers using smart cards tend to prefer a full accounting of the goods and services they purchase. When the card is used to calculate road tolls, people are often anxious to make sure they have not been shortchanged.[61] One US assessment of public responses to road-toll technology observed:

Concerns that motorists would feel their privacy was compromised under an ETTM [Electronic Toll and Traffic Management] system which recorded vehicle movements prove to be unfounded. Toll agencies that record this information do not make it available to outside parties. In fact, existing ETTM experience reveals that a large majority of motorists choose payment options (often via credit card) which do not provide anonymous transactions.[62]

Privacy Rights Are Becoming Commodities

Privacy's journey from the political realm to the consumer realm invokes a new set of relationships and values. Placing privacy in the "free market" environment along with such choices as color, durability, and size creates an environment in which privacy becomes a costly "add-on."

The debate over Caller ID highlights this trend. The status quo for 80 years has been for the option of disclosure by the calling party. In many parts of the world, this technology has not only reversed the onus of disclosure; it has also imposed financial and other costs on those who wish to maintain their traditional level of privacy.

The process of commodification is inimical to privacy. Every element of privacy protection is interpreted and promoted as a direct cost to the consumer. The cost factor is a powerful weapon in the armory of privacy invaders because it implies that a few "fundamentalists" will force a rise in the production cost of an item or a service.

The Triumph of "Public Interest"

The concept of public interest appears to have expanded substantially in recent years, to the point where in some countries it is now a generic description for all activities of the state. This can be explained in part by the overall level of anxiety about threats such as crime and unemployment. It is also likely that "public interest" can be manufactured limitlessly in any environment in which political choice is minimal (as it was in East Germany and other Eastern Bloc countries).

There may be a perception in the public mind that solutions to problems can be found only through institutions. The division between private and public organizations, between the surveyors and the surveilled, between the individual and the state, is fast disappearing.

Conclusion

The upsurge in surveillance by private and government bodies around the world is closely linked to the redefinition of privacy, to the resort to fixed data-protection principles, and to a massive expansion in the notion of "public interest."

Although it is natural to avoid any hint of conspiracy theory in this scenario, it is hard to avoid the conclusion that information users have reached a degree of common ground in their approach to the privacy issue. Common ideas and expressions emerge everywhere in the world.

From the perspective of privacy reform, the current trends are matters of deep concern. Though it is true that the concept of privacy has become fragmented and dispersed among many sectors and interest groups (medical privacy, genetic privacy, etc.), the loss of traditional privacy activism at a macro political level has imperiled an important facet of civil rights.

Notes

1. Cited in Simon Davies, *Big Brother* (Pan Books, 1996), p. 53. Harmony ideology is described as the coming together of opposing ideologies and beliefs into a manufactured consensus.

2. David H. Flaherty, *Protecting Privacy in Surveillance Societies* (University of North Carolina Press, 1979), p. 374.

3. D. Campbell and S. Connor, *On the Record* (Michael Joseph, 1986), p. 20.

4. The constitutions of Germany, Hungary, and Portugal outlaw national numbering systems, which they consider dangerous to human rights. The Scandinavian countries have enshrined very strong freedom-of-information laws. France maintains the right to ban hostile technologies.

5. Alan F. Westin, *Privacy and Freedom* (Atheneum, 1967), p. vii.

6. The most recent attempt to pass general privacy legislation was a 1994 bill presented by Senator Paul Simon (Democrat, Illinois). This failed to attract even enough support to take the form of an amendment to other legislation.

7. *EPIC Alert* 2.14, November 9, 1995. (*EPIC Alert* is an electronic bulletin published by the Electronic Privacy Information Center in Washington.)

8. Ibid.

9. Flaherty, *Protecting Privacy in Surveillance Societies*, p. 7.

10. Australian Privacy Commissioner, Community Attitudes to Privacy, Information Paper 3, Human Rights and Equal Opportunity Commission, Sydney, 1995.

11. "Public interest" can be loosely defined as any exemption from a legally protected individual right. In the case of personal privacy rights, public interest often includes public health, national defense, the economic well-being of the state, national security, activities of law enforcement agencies, and public security. The preamble of the European Data Protection Directive also cites examples of international transfers of data between tax or customs administrations or between services competent for social security matters as constituting public interest.

12. Will Hutton, *The State We're In* (Vintage, 1996).

13. Lance Morrow, The temping of America, *Time*, March 29, 1993.

14. J. Rifkin, *The End of Work* (Putnam, 1995), p. 190.

15. Learning to cope, *The Economist*, April 6, 1996, p. 15.

16. C. Negrey, *Gender, Time, and Reduced Work* (State University of New York Press, 1993), p. 43.

17. World Labor Report, 1993 (International Labor Office, Geneva), pp. 65–70.

18. Hugh Mackay, *Reinventing Australia* (Angus and Robertson, 1993), p. 19.

19. The number of notifiable offenses recorded in England and Wales has consistently declined since 1993 (Home Office Statistical Bulletin, March, 1996), while responses to crime surveys indicate increasing levels of fear about crime. One Home Office survey of Birmingham residents indicated that 45% felt "very unsafe" after dark (CCTV in town centers: Three case studies, Police Research Group, Crime Detection and Prevention Services, paper 68, 1995). For a discussion of the impact of media reporting on fear of crime, see J. Lovering, Creating discourses rather than jobs: The crisis in the cities and the transition fantasies of intellectuals and policy makers, in *Managing Cities: The New Urban Context*, ed. P. Healy et al. (Belhaven/Wiley, 1995).

20. S. Flusty, Building Paranoia: The Proliferation of Interdictory Space and the Erosion of Spatial Justice, Los Angeles Forum for Architecture and Urban Design.

21. Ibid.

22. These systems were pioneered in the 1970s at Scottish defense establishments, such as the Faslane submarine base.

23. Simon Davies, *Big Brother* (Pan Books, 1996), p. 183.

24. Ibid., p. 183.

25. Use of police cameras sparks rights debate in Britain, *Washington Post*, August 8, 1994.

26. CCTV briefing session organized by Computer Security Research Centre, London School of Economics, June 5, 1996.

27. Stephen Graham, John Brooks, and Dan Heery, Towns on the Television: Closed Circuit TV in British Towns and Cities, Centre for Urban Technology, University of Newcastle upon Tyne.

28. Davies, *Big Brother*, p. 182.

29. Ibid., p. 177.

30. CCTV, Looking Out for You, Home Office, London, 1994, p. 12.

31. In *Big Brother* I cite numerous examples of such schemes as ID cards, numbering systems and community information gathering schemes that rely on such thinking.

32. Terry Honess et al., Closed Circuit Television in Public Places, Police Research Group, Home Office, 1992.

33. "20/20," ABC news, September 7, 1995.

34. S. Davies, They've got an eye on you, *The Independent*, November 1995.

35. CCTV Code of Practice, Local Government Information Unit, London, 1996.

36. For a discussion of these points see S. Davies, *Monitor: Extinguishing Privacy on the Information Superhighway* (Pan Books, 1996).

37. Ibid., p. 27.

38. Westin, *Privacy and Freedom*, p. x.

39. In 1983 the residents of Germany responded to their national census in the same fashion, and this resulted in a landmark Constitutional Court decision protecting personal information rights. See Flaherty, *Protecting Privacy in Surveillance Societies*, pp. 45–46.

40. The Schengen Agreement on police cooperation, which currently involves about half the countries in Europe, was designed to ensure that the dismantling of border controls did not happen at the expense of public and national security. The result was a strengthening of the powers of national police authorities, and the development of the Schengen Information System to share data among countries. For a detailed discussion of the Schengen Agreement and its implications, see various issues of *Statewatch*.

41. S. Davies, Touching Big Brother: How biometric technology will fuse flesh and machine, *Information Technology and People* 7 (1994), no. 4, p. 43.

42. The CSA standard has been designed as a measurable privacy code. It is the first such privacy standard.

43. An inaugural public meeting was held in Auckland on September 13, 1991.

44. Davies, *Monitor*, pp. 154–155.

45. EPIC was formed in 1993 as a non-government watchdog over threats arising from electronic surveillance and censorship, and to champion a range of consumer, freedom-of-information, and privacy issues. The group has also sponsored Privacy International, the sole global non-government body in this field.

46. In early 1994, EPIC and Computer Professionals for Social Responsibility launched an electronic petition that resulted in more than 50,000 signatures against the Clipper encryption proposals.

47. In July 1996 an anonymous web site was established to convey "real-time" images from a camera placed outside the home of Defense Secretary Michael Portillo. The tactic, and the controversy which followed, succeeded in highlighting a contradiction in the frequently expressed government view that no right of privacy exists in a public space.

48. EPIC succeeded in 1993 in reversing a plan by Lotus to introduce a CD-ROM containing details of most American households. It did this by posting the personal e-mail address of Lotus's chief executive officer on the Internet, and announcing an e-mail campaign. The CEO received more than 30,000 messages opposing the plan, and the CD-ROM was never released.

49. Christopher Lasch, *The True and Only Heaven* (Norton, 1991); *Resisting the Virtual Life*, ed. J. Brook and I. Boal (City Lights, 1995); Edward Tenner, *Why Things Bite Back: New Technology and the Revenge Effect* (Fourth Estate, 1996).

50. Jan Holvast, Vulnerability of information society, in *Managing Information Technology's Organisational Impact*, ed. R. Clarke and J. Cameron (North-Holland, 1991).

51. Privacy International, interim report to members, Sydney, 1991, p. iv.

52. David Flaherty, The emergence of surveillance societies in the western world: Toward the year 2000, *Government Information Quarterly* 5 (1988), no. 4: 377–387.

53. Flaherty, *Protecting Privacy in Surveillance Societies*, p. 385.

54. Stakeholder analysis, a popular business model for some years, is particularly relevant to interorganizational relationships. For an overview of the approach see W. N. Evan and R. E. Freeman, A stakeholder theory of the modern corporation, in *Ethical Theory and Business*, fourth edition, ed. T. Beauchamp and N. Bowie (Prentice-Hall, 1993).

55. A campaign of opposition was started by the Australian Doctors' Fund, but opposition to the proposed network soon spread to the more conservative Australian Medical Association.

56. *Daily Telegraph*, London, April 4, 1995.

57. Ibid.

58. David Hencke, Howard's ID smart card plans found in junk shop, *The Guardian*, January 16, 1995, p. 1.

59. Davies, *Big Brother*, p. 134.

60. Ibid., pp. 69–72.

61. Ibid., p. 172.

62. Neil D. Schuster, ETTM technology: Current success and future potential, in Proceedings of the IVHS America 1994 Annual Meeting.

6

Controlling Surveillance: Can Privacy Protection Be Made Effective?

David H. Flaherty

There is wide-ranging theoretical and empirical support for the importance of personal privacy in our respective lives, but there is some disagreement among nation states about how best to implement the protection of privacy of individuals in the public and private sectors. Instead of simply thinking about the issue and trying to develop some kind of theory about how best to proceed, my interest during the last 20 years has been to observe the kinds of significant privacy problems that have emerged in advanced industrial societies and then seek to monitor, through case studies, the various approaches of "regulatory" bodies, with widely varying powers, as they try to fashion solutions for the challenges to individual privacy that continue to surface.

Most of what I have learned was published in my 1989 book *Protecting Privacy in Surveillance Societies*,[1] which examined the workings of national and state data-protection agencies in the Federal Republic of Germany, Sweden, France, Canada, and the United States. As I am both an academic (at the University of Western Ontario) and a consultant on privacy policy, my findings and recommendations were very much shaped by direct observation of data protectors at work in many countries and by an examination of what they wrote, and what was said and written about them, in scholarly and journalistic circles. It is an understatement to say that examining the functioning of such small, important bureaucracies was a significant learning experience for me. But I do not think that it ever came into my consciousness that I would be given an opportunity to do such work myself in Canada.

In the 1990s I became interested in trying my hand at implementing privacy protection officially. Canada has had the post of federal Privacy

Commissioner since 1977; this post is filled through a political process dominated by the prime minister's office. Some of the provinces have similar posts. Quebec has had a Commission on Access to Information since 1982, but its members are also selected by the government of the day in what I would also describe as a political process.

In 1988, Ontario advertised for an Information and Privacy Commissioner and then allowed a select committee of the legislature to make the choice, subject to legislative approval. In 1991, when this post was open, I was not selected. In July 1993, however, also on the basis of a public competition, I became the first Information and Privacy Commissioner for the Province of British Columbia, which has a population of just under 4 million and is the third largest of the Canadian provinces (after Ontario and Quebec).

My provincial counterparts in Quebec, Ontario, and Alberta are both Privacy Commissioners and Freedom of Information Commissioners. These positions are still separate at the federal level, although there has been misguided agitation for their integration. (In my view, the sheer physical expanse of Canada necessitates separate commissioners to promote these competing interests effectively, especially since the staffing and budgets of the federal offices are comparatively small.) Canada is in fact unique in formulating laws and enforcement mechanisms, together, both for greater openness and accountability for general information in society and for the protection of the personal information of the citizenry. In the first instance, this process emulates of Sweden and the United States, but, especially compared to the latter, Canada is much more advanced on the implementation side of the ledger; that is, there is someone in charge of making the law work for citizens. Americans usually have to sue in the courts to achieve access to information or protection of their privacy when they encounter obstacles or practices that they consider objectionable.

In this chapter I propose to ignore, for the most part, the freedom-of-information side of my work in British Columbia, which involves the exercise of regulatory power in decision making on specific cases, and to concentrate on what primarily interested me in the writing of my 1989 book: how to try to control surveillance by making data protection effective in the public sector. Although I am more than three years into a six-year mandate as Commissioner, I want to reexamine the conclusions to my book by testing my academic findings against my practical experience. I do so fully cognizant of the risks of egocentrism's being the primary theme of what follows. I hope that by being self-critical and by giving fair play to my critics I can avoid wallowing in excessive self-admiration. I also acknowledge, in advance, that any model of data protection is much affected by the personality of the commissioner, the political culture of a particular jurisdiction, and the state of the local economy. Good times are better for data protection than bad times.

I would suggest that a cynical legislature, wanting to offer only the illusion of data protection, should not in the future select a privacy commissioner who knows anything about his or her duties at the time of selection, since there is a significant learning curve for any new appointee, especially if the office is new. Based on my 20 years of research and consulting activities concerning the protection of privacy, I knew in particular that I had to adopt a proactive approach to the advocacy side of my work (the "privacy watchdog" role), which continues to be a big surprise to politicians and to the bureaucrats who actually run government in the trenches. "What is Flaherty trying to do to us?" has been a not-uncommon response by the government in power and by the public service to my advocacy role and my media statements on privacy issues. Just as the bureaucracy is having to learn to live with freedom of information and with privacy protection, so we have been learning how to live with one another and to appreciate our varying responsibilities. I have not suffered from a lack of invited and volunteer tutors.

The Need for Data-Protection Laws

As I noted in *Protecting Privacy in Surveillance Societies*, "the existence of data protection laws gives some hope that the twenty-first century will not be a world in which personal privacy has been fully eroded by the forces promoting surveillance."[2] This debate has been handily won by the pro-privacy side of the equation, at least in most advanced industrial societies. It is now an odd Western nation, state, or province that does not have explicit laws for controlling the collection, use, retention, and disclosure of personal information in both the public and the private sector.

In the 1980s I was still somewhat cautious about fully endorsing the need for data-protection laws. The deregulation movement led by Margaret Thatcher and Ronald Reagan was in full flood, and the American resistance to regulatory intervention of this type was also flourishing (with interventionism being ahistorically viewed as some kind of un-American activity, a kind of first step toward a socialist state). But advances in the development and application of information technology since the mid 1980s, almost bewildering in their scope and pace, have lent considerable urgency to the argument that government's and the private sector's rapid adoption and application of numerous forms of surveillance technology require a watchdog mechanism and a set of statutory rules for fair information practices. Data matching, photo radar, digital photographs, criminal-record checks, and pharmacy prescription profiling systems are some local examples that spring to mind. As I wrote in 1989, "the critical issue is how best to strengthen data protection in the face of strong, sustained, countervailing pressures for surveillance."[3]

In fact, my experience in British Columbia is that the pressures for surveillance are almost irresistible for an office such as mine. The bureaucrats and the legislature are under intense pressures to reduce costs, to promote efficiency, and to spend public money wisely. Surveillance technology appears to be a neutral, objective process that must be wielded as a weapon, or at least a tool, against welfare cheats (targeting all those on income assistance), sex offenders (targeting all those who work with children through criminal-record checks), and photo radar (monitoring all cars and photographing the license plates of speeders).

I also notice considerable pressure in British Columbia for the rationalization of identity checking of applicants for government benefits, whether for income assistance, health care, or drivers' licenses, on the basis of what is euphemistically called a "common registry." This registry appears to be the first step toward the development of a national data bank, which has figured in the nightmares of US privacy advocates since the 1960s. The fact that such technological applications are so apparently achievable drives these seeming imperatives among legislators, functionaries, and the taxpaying public. One consequence will be a full-fledged surveillance state. I really fear that such a result is almost impossible to prevent in the long term, whatever the prodigious efforts of data protectors and their allies.

Thus, my stated ambition in 1989 of "protecting the personal privacy of individuals to the fullest extent possible in a democratic society"[4] seems to be increasingly problematic. Neither government nor the private sector really likes the privacy business, whatever it is, because it gets in the way of their continuing to do business as usual with personal information. (Here I am referring to the government as a collectivity rather than as specific individuals with human concerns: I find it tedious to be told, repeatedly, by those practicing the systemic invasion of privacy how much they value privacy in their own lives. Getting government officials to think as human beings, with families and friends, is an ongoing challenge for data protectors.)

Concern for privacy comes to be perceived as an unnecessary barrier to what is regarded by most legislators in governing parties, and by some taxpayers, as clearly rational behavior. Hence the value, in my opinion, of having data-protection laws and officials in place. They are at least a temporary barrier to what others naturally perceive as progress in dealing with some problem of the moment that seems to require fixing. The "fix-it" solution is even more software, systems, record linkages, and advanced surveillance technology. These trends place data protectors in the awkward and unpleasant position of at least appearing to resist multiple forms of progress—but that is what they and their staffs are paid to do.

Defining Privacy Interests to Limit Surveillance

It is no particular surprise that legislators in British Columbia, as elsewhere, did not heed my advice to define privacy interests more adequately in legislation as a guide to limiting surveillance.[5] I conclude that those who draft legislation must in fact throw up their hands in bewilderment at such a prospect, if they have indeed ever been invited to tackle the thorny issue of defining privacy. Only philosophers continue to bemuse themselves with this important activity, it would appear, while individual authors parade their ingenuity with increasingly obscure, and obscuring, definitions.[6]

I have had no more success at focusing legislative efforts on articulating the central privacy interests that require protection. However, let me say, immodestly, that among the most useful features of my book are several tables that seek to summarize the "privacy interests of individuals in information about themselves"[7] and "data protection principles and practices for government information systems."[8] The latter table originated as an *aide memoire* as I attempted to track the most appealing features of extant laws and major legal decisions. A number of them, including the principles of informational self-determination, publicity, transparency, finality, data trespass, and the right to be forgotten (all admirable products of European thinking and lawmaking), do not appear explicitly in the BC Freedom of Information and Protection of Privacy Act. My conscious goal has been to promote these principles and practices on the local scene by reprinting the two tables in the introduction to my first annual report for 1993–94, referring to them in our Investigative Report No. P95-005, March 31, 1995, and generally discussing these broad measures to protect privacy in my ongoing media relations and public education efforts in the province. The "right to be forgotten" figured in one of my recent orders (Order No. 58-1995, October 12, 1995, p. 7).

I wrote in 1989 that "protection of privacy should also remain largely a nonlegal activity in most areas of human existence, meaning that even when data-protection laws exist, individuals have to rely to a considerable extent on their own efforts, such as refusing to give out personal information, in order to limit unwanted surveillance."[9] In old and new ways, data protection, in my considered view, remains largely a matter of raising consciousness at various levels of society. Despite my ongoing efforts to act as the privacy watchdog for the province, I regularly end my interactions with the general public, on radio talk shows in particular, by encouraging every listener to be their own privacy watchdog by controlling what they choose to disclose to outsiders and by monitoring what they are being asked to disclose. More and more people appear to be prepared to say that a request for their personal information is none of anyone's business. A more privacy-conscious public service means that privacy-impact statements for proposed legislation are prepared at an appropriate time in the annual legislative calendar without my office's always having to intervene to stimulate such an activity in the first place. The several times that this has happened in British Columbia, especially through the good offices of the Ministry of Health, are an ongoing sign of progress in data protection.

In classic liberal fashion, I emphasized in my book the need to balance privacy against competing values and legitimate governmental purposes.[10] My efforts to appear balanced to consumers of my research strike me as naive in retrospect: the striking of balance within government is so much against the privacy interests of individuals that it is a wonder we have any privacy left once governments get through doing what is good for each and every one of us. What is good for government is always thought by those in government to be good for the public at large.

I emphasized in 1989 the importance of minimizing data collection in order to reduce intrusion into private lives.[11] I am more than ever convinced that this is a central concept in debates over privacy and its preservation. Personal information must be collected and shared only if there is a demonstrated need for it in the first place. As a "culture of privacy" begins to impinge more on the operating consciousness of the hundreds of thousands of persons working for public bodies subject to the BC Freedom of Information and Protection of Privacy Act, I find modest evidence of a self-initiated process of questioning why certain data should be collected. Among other things, we have succeeded in ending the collection of the Social Insurance Number, a unique personal identifier, in at least a dozen settings, including its use as a customer control number by BC Hydro.[12] On the other hand, we have also accepted the explanations of the BC Gaming Commission as to why it feels compelled to collect so much personal information about the people associated with applications for licenses to run gambling establishments.

The Need for Data-Protection Agencies

In 1992 the legislature of British Columbia, led by a newly elected New Democratic Party, created the Office of the Information and Privacy Commissioner in imitation of the Ontario and Quebec models. Alberta, under a very conservative premier who has substantially cut back government services, imitated British Columbia in 1995. Thus, all the most populous provinces in Canada now have data-protection agencies.

I need to separate my support for such initiatives, both in theory and in practice, from my criticisms of how privacy commissioners actually go about their work, since these are quite different matters. In my judgment, the best evidence of the need for privacy-protection bodies is the sorry state of data protection in the US, where only a few of the states have equivalents of the Canadian agencies. Those brave souls in New York, Connecticut, Minnesota, and Hawaii work valiantly within the executive branch to promote freedom of information and privacy, but they lack independence, financial and human resources, and political clout. What they have accomplished gives some sense of what is needed and of what is possible in a market-driven capitalistic society that treats personal information as simply one more tradable commodity.

The United States suffers from a lack of an ongoing, institutionalized body of specialized privacy expertise directly available to all levels of government. The small network of talented US privacy advocates, including lawyers in private practice, cannot compensate adequately for this lacuna. It is demoralizing to watch admirable organizations like the Electronic Privacy Information Center (EPIC) have to bring lawsuits to accomplish privacy-protection goals that are rarely recognized in the federal and state legislative arenas. Although there is a clear role for litigation in the implementation model for any such law, it should not have to be the first recourse for concerned or affected citizens, who can rarely afford such a luxury as suing the government.

Although my office has encountered resistance to its data-protection work, one of our considerable accomplishments in a geographically limited area has been to bring our existence to the direct attention of policymakers, senior politicians, policy analysts, and other public servants, at least in the central government. Whatever they may think of our actual work, they are aware that we exist. My proactive advocacy role on privacy issues has surprised them, I think, because the provincial Ombudsman and Auditor General function somewhat differently than the commissioner and because neither of them has regulatory power. They may identify problems, and they may make recommendations; however, because I have the authority to make binding decisions in access-to-information cases, my "recommendations" carry considerable weight in achieving compromise solutions on privacy policy (unless, of course, the legislature decides to move ahead with its plans anyway—e.g., by implementing a province-wide system of criminal-record checks to uncover pedophiles with criminal records). But I am not part of the hierarchy of public service, and some senior bureaucrats have trouble accepting that I cannot be controlled, in the traditional sense, for the sake of specific political or administrative agendas (even though informal pressures of that sort exist).

I wrote in 1989 that data protection "requires significant modification in information-handling practices."[13] I also thought that data protectors always encountered significant resistance from one organization or another. I can honestly say that I have not met or felt such overt resistance, especially on the privacy side of my work, although this may be due to a lack of perception or of sensitivity on my part. The most significant resistance has come from the Ministry of Human Resources, which has an enthusiasm for data matching. We have not found examples of significant bad actors who require numerous remonstrances; perhaps the "bad guys" are still hiding from us. This presumably happy situation is explained in part by the fact that we are still in the honeymoon phase of our relationship with government and in part by BC's relatively robust economy. Budget cutbacks, which are now beginning, will have an impact on the capacity of my office to accomplish its statutory tasks. We will have to concentrate on our core activities, which include mediating requests for review of access requests, investigating complaints, giving policy advice to public bodies, conducting site visits as a form of auditing, and engaging in media relations and public education.

I still like my 1989 formulation that "data protection commissioners or agencies are an alarm system for the protection of privacy."[14] Our tools include oversight, auditing, monitoring, evaluation, expert knowledge, mediation, dispute resolution, and the balancing of competing interests.[15] Thus, I continue to believe that "data protectors are specialists in articulating privacy interests on a systematic basis on behalf of the public and in applying fair information practices."[16]

The risk side of the equation is that we will "simply become agents for legitimating information-collection activities and new information technology."[17] We may act only "as shapers of marginal changes in the operating rules"[18] for instruments of public surveillance, losing battles as often as we win them. I certainly feel strong pressure to get along with my central-government colleagues, and my staffers are all public servants who might like to work elsewhere in government later in their careers.

To date I do not deserve to be considered an active agent for the legitimization of government practices, although I have had my failures. At the end of two years of discussions, the Pharmanet prescription profile system went into effect with most of my suggestions for data protection incorporated , but I could not stop the mandatory nature of the system in the face of the cabinet's desire to establish it as a cost-saving measure. I had no impact on another massive system for invading the privacy of one-sixth of the adult population, criminal-record checks for all those who work with children—the government and the legislature were not persuaded by any of my private or public advice. I had to accept a system of data matching intended to ensure that British Columbia's recipients of income assistance are not also receiving such help from other provinces. (The record of cheating is undeniable. The privacy of all those on income assistance is invaded because of the fraudulent behavior of a small minority. We succeeded in getting a requirement that all recipients of income assistance be notified of the data matching, and in getting ongoing monitoring of the fraud-prevention process. When we questioned the legal authority of the Ministry of Human Resources to perform data matching, the ministry secured legislative authority for the activity, thus minimizing my impact and authority.)

The Effective Conduct of Data Protection

In compliance with my own admonitions, I have adopted and applied a functional, expansive, and empirical, rather than a formal and legalistic, approach to my statutory tasks. In essence, I try to keep the broad goals of protecting privacy in mind rather than worrying unnecessarily about legal niceties. I do have to follow the Freedom of Information and Protection of Privacy Act; however, as I said in my Order 74-1995, I am not much interested in making Jesuitical distinctions that would bring credit to a medieval theologian but would not make much sense to the taxpayers of British Columbia.

As a matter of fact, I find myself preoccupied in more than 140 decisions to date with the broad goals of the aforementioned act. My greatest goal has been promoting openness and accountability of government to the public. My Achilles' heel with respect to judicial review is proving to be an expansive application of solicitor-client privilege, despite my efforts to restrict its application in accordance with the broad goals of the act. With respect to the privacy side of my dual mandate, and this may be an inherent limitation of the single-commissioner model, I can be accused of having occasionally sacrificed the privacy interests of individuals in the pursuit of greater openness.

Fortunately, section 22 of the act contains a set of privacy standards that my colleagues and I apply to specific decisions about the disclosure of personal information by public bodies. This has involved very nice distinctions in about half of my decisions. I, like others, will be quite interested to see where the bright lines are finally drawn in this regard about such problems as disclosing the identities of complainants or informants. I believe that my own emphasis on making decisions on a case-by-case basis, rather than worrying too much about consistency, fits into a pragmatic and functional approach to my statutory tasks. I am clearly concerned about making appropriate policy decisions on the basis of the act rather than being what Americans call a strict constructionist.

The empirical approach of my office also deserves attention. We are interested in knowing, in as much detail as is necessary, how a proposed new system or a modification of an existing one will actually work. This is not a trite point, because I believe it is an extremely important perspective for a data protector to adopt. One cannot regulate a system without fully understanding it.

Another important component of our approach is avoiding public and private confrontations by being pragmatic in our approach to negotiations. Spiros Simitis (the data-protection commissioner of the German state of Hesse) wisely described this as "a gradualist or incremental approach to data protection, involving cooperation, a process of learning, and a search for consensus."[19] We are not even remotely interested in fighting for the sake of fighting. I also fully agree with Simitis, based on practical experience, that our work is political in the best sense of that abused word, which means, in this context, having good working relationships with those being regulated in order to convince them that they are being treated fairly, seeking appropriate allies as needed, and communicating effectively with the public through the media.[20]

I can only smile, diffidently, at my description of the ideal data-protection commissioner as a person who is "self-confident, perceptive, experienced, well-connected, reasonable but firm, has a strong presence, and is politically astute."[21] In terms of building a constituency, I have been blessed with the lobbying and public-education activities of the Freedom of Information and Privacy Association (FIPA) and the British Columbia Civil Liberties Association (BCCLA). The concentration of perhaps two-thirds of the population of British Columbia in greater Vancouver and Victoria makes it easier to reach the public and to interest the media in the work of our office. In terms of promoting my stated views on how things ought to be done in order to make data protection effective, I have been singularly blessed by being educated at the feet of some of the great practitioners of the art, including Spiros Simitis, Hans Peter Bull, Jan Freese, and John Grace. Thus, it would have indeed been surprising if I did not know what I should at least try to do in the conduct of this "vital activity."[22]

I especially like being reminded by my own writings that it is necessary to understand and be mindful of the competing interests of the government, the legislature, the public service, and various segments of the public.[23] There is no use pretending that privacy protection is the main purpose or activity of any part of government. Competing interests must be taken account of in finalizing decisions on specific data-protection matters and in finally deciding what advice to offer a public body about a particular program or practice. My oversight of certain aspects of the work of municipal police forces is a good case in point. I have been concerned with such issues as the security of records and the adequacy of audit trails for police users of the Canadian Police Information Centre (CPIC).

In my 1989 book I vacillated between the avoidance of bureaucracy in a data-protection organization and the risk of having "an overworked and overcommitted staff that provides only the illusion of data protection."[24] My office has not had the time to become either overworked or overcommitted, although I plead guilty to the common sin of neglecting inspections or audits (which I prefer to call site visits). This activity takes second place to my particular need to write orders to settle specific cases.

I fully agree with my prescriptive statement that the primary role of a data-protection agency is the actual articulation and advancement of the privacy interests that must be defended in a particular setting.[25] In my case, the public body and/or the legislature eventually establishes what they perceive to be the acceptable balance among competing interests (not always to my complete satisfaction). For specific cases, however, I reach conclusions about appropriate information practices as the decision maker, subject to judicial review in the courts.

Independence and the Exercise of Power

The exercise of independence by a regulatory authority, subject to appropriate external controls, is vital to successful data protection. In British Columbia, I am somewhat protected as an officer of the legislature, with a fixed six-year term that is not renewable, a salary tied to that of the chief judge of the Provincial Court, and a generous pension plan for the actual period of employment. Because I knew from my studies about the importance of the independence factor, and because of my own personality, I have emphasized my independence in words, actions, and orders. I am protected by having a tenured professorship to return to at the end of my appointment, the importance of which I knew from the previous experience of Hans Peter Bull and Spiros Simitis in Germany.[26]

The fact that my budget ($2.6 million Canadian) comes from the same Treasury Board that funds the entire public service has not escaped my attention. Fortunately, I have an approved staff of 25. That is the number that I planned for from the beginning, and I am committed to accomplishing what we can with that staff by learning to work more and more effectively. But I am also subject to the same budget cutbacks that have occurred throughout the government of British Columbia. If our workload continues to grow significantly, I may also find myself with cap in hand before the Treasury Board, seeking to justify increases in budget and personnel in a very tight fiscal situation.

My staff consists of public servants appointed in compliance with the Public Service Act, which at least makes them insiders to government in terms of potential job mobility. Because of the wishes of the legislature, none of them are union members.

My concern for independence is counterbalanced by the desire to build an effective network in government circles to facilitate the mediatory role of our office in settling most of the requests for reviews of access decisions that come to us. While I know that networking is necessary, there is also a risk of our being coopted by a desire to get along and go along. Again, I am very grateful that two public-interest advocacy groups, the Freedom of Information and Privacy Association and the BC Civil Liberties Association, keep our feet to the fire with constructive criticism.

Fears among public bodies of my "independence" may also have been mitigated by the fact that they have to date been successful in about two-thirds of the cases I have had to decide. While my decisions against the government tend to receive a lot of publicity, especially from successful media applicants, public bodies have won major victories on attempts by the public to access long-distance telephone logs and back-up e-mail tapes. I also like to emphasize that my 140 decisions to date have been reasonable and pragmatic, hoping thereby to assuage the ongoing anxieties of more than 2,000 public bodies by demonstrating that the intervention of my office in their routine work is well considered, with appropriate deference to professional judgment and awareness of the practical realities of running a government or providing public services.

My own office lacks an oversight mechanism for reporting to the legislature in a truly meaningful way. Receipt of my annual report is a relatively perfunctory matter; it is simply tabled with the reports of everyone else, and no legislative committee discusses it with me. The Privacy Commissioner of Canada and my counterpart in Ontario appear to have somewhat more meaningful reporting relationships with their respective legislatures. My annual reports have been relatively nice to government and not especially controversial. I have also not issued a special report to the BC legislature on a particular problem; that would likely attract more attention.

The flip side of this issue of reporting is that my situation also prevents direct interference by legislators in my work. I did have the experience of being told publicly by one member of the legislature that I worked for her. I prefer, now, to state that I am appointed with a six-year no-cut contract, subject to impeachment for maladministration and other heinous matters. There is a provision for a special committee of the legislature to begin a review of the functioning of the act within 4 years of its going into effect (meaning October 4, 1997), and I expect at that time to revisit this matter.

My main contacts with members of the legislature, especially with members of the select committee that appointed me in 1993, are on an individual basis. As a result of my negative experience with criminal-record checks in the spring of 1995, I am committed to directly informing the 75 members of the legislature about my views on privacy matters coming before them.

The legislature has amended the Freedom of Information and Protection of Privacy Act on several occasions, but not in ways that affect my work. The major amendments have been "notwithstanding" clauses to other acts to protect the operations of certain specialized agencies of government on freedom-of-information matters. My advice to the cabinet on most of these proposals was accepted. How to provide for adequate warnings of future contemplated changes is problematic; again, networking appears to be the solution.

With respect to the exercise of power by my office, I have heard vague rumblings that I have too much power and related complaints that I am making policy for the government. Though we have not had "an uphill fight" to make our voices heard,[27] I am more aware than at the time of my appointment of the limits of what my office can accomplish, even with the expansive mandate entrusted to it under section 42 (the "information czar" clause) of the act. We have the resources to be at most a thorn in the side of thousands of public bodies. We are also having to continue to work hard "to win the cooperation of the regulated,"[28] a task admirably performed by a director, a dozen portfolio officers, and two intake officers in my office.

Our main problem is learning about the major initiatives of public bodies at all levels that have surveillance implications when it is still possible to do something about them, although I think that my colleagues and I now have better antennae and a better network than we did in our early days. The government assigns task forces and study teams to examine various initiatives and then sometimes refuses, or is reluctant, to tell me about them because the cabinet has yet to give its approval. But once the cabinet has acted, my hands are relatively tied. There are also so many ongoing issues of access to information and data protection that crisis management is almost the dominant mode of operating. And the more orders I produce, the more one public body or another may have reason to be unhappy with me. They remember their losses much more acutely than their victories.

At the end of the day, as I have noted above, my independence and power are considerably limited, in the policy field, by the authority of the legislature to make any statutory decision that it wishes to introduce or to increase surveillance of the population in some manner.[29]

I concluded my 1989 treatment of these issues by emphasizing the importance of personnel selection and public relations. I have been most fortunate to recruit a professional staff of considerable experience and varied professional and administrative backgrounds and careers. It has been difficult for anyone with less than 10 years' experience to win a posting at the senior level in my office. These individuals not only know how government works, they also fully appreciate our independent role. Several key personnel came from the office of the ombudsman and thus understand intimately the role of guardian agencies in the political and bureaucratic process.

The Adequacy of Advisory Powers

This is the area where my views have changed the most since 1989 as a result of my direct experience with a regulatory data-protection statute in British Columbia. I only hinted then at the fact that advisory powers might not be adequate to the tasks at hand.[30] I now believe that a privacy commissioner should have regulatory power at his or her disposal. Yet the reality is that most of the problems brought to my office are settled through mediation, not through orders and binding rulings. Our approach is very conciliatory. We act "in a flexible and pragmatic fashion, responding as much as possible to experience and learning" from our mistakes.[31]

Our work has been facilitated by considerable media coverage. It is difficult to measure what the public thinks about us, but those in the Lower Mainland and in Victoria have been given many opportunities to learn about our watchdog role in privacy matters. Since the media tend to support decisions favoring their requests for access to information tends, there is probably a "halo effect" transferable from my work as Information Commissioner to my work as Privacy Commissioner. Independent surveys of public opinion continue to document very high levels of concern in British Columbia for the protection of individual privacy: "Outbreaks of aroused public concern create a positive climate for the implementation of data protection."[32] A series of relatively minor privacy disasters, especially in the health and medical field, have been very important for consciousness raising among the general public. Sensitive personal records being faxed to the wrong places have been prime examples of this trend. This led the Ministry of Health to authorize a very useful report by Dr. Shaun Peck, the Deputy Chief Medical Officer for the province, entitled Review of the Storage and Disposal of Health Care Records in British Columbia (1995).

One "weakness" in the Freedom of Information and Protection of Privacy Act that I anticipated in 1989 is the absence of statutory sanctions. If a public body or its agency fails to comply with the fair-information-practices provisions of the act, there are no criminal sanctions in the act itself, however serious the breach. I am satisfied for the moment, however, to rely on progressive disciplinary proceedings for the exercise of social control. But a four-day suspension of a municipal police officer for unauthorized use of the Canadian Police Information Centre to access the provincial motor-vehicle database was widely perceived as too small a slap on the wrist, and indeed a complainant has appealed it. While other "offenders" might be prosecuted under various sections of the Criminal Code of Canada, this may not be enough to satisfy the public lust for blood when a serious breach of the act occurs. I maintain a holding brief on this matter as I await guidance from direct experience. I sense that the public would like to know that there is a provision in the act to "punish" someone when egregious breaches of confidentiality occur.[33]

The Primacy of Data-Protection Concerns

I argued in 1989 that privacy protectors should stick to their knitting and avoid direct responsibility for the other information-policy issues of their respective societies. The implication was that they should avoid issues not related to controlling surveillance of the population and emphasize the protection of individual human rights.[34] The pressures to meddle in broad issues of information policy are indeed great, especially as the "information highway" debate rolls along and especially for an individual who is both an information commissioner and a privacy commissioner. As I have said above, I am aware of the limits of what my office can accomplish with the resources that we can legitimately expect society to entrust to us. What I have done is establish working relationships with others charged with such matters as promoting computer security, including people in the office of the province's new Chief Information Officer. I very much welcome the fact that the Chief Information Officer has assumed overall responsibility for the province's information policy. I like to be kept informed of important developments affecting information policy, but there are limits to what I can do to shape them. On the other hand, I prepared submissions on privacy issues to the Industry Canada task force on the information highway and to similar deliberations on the same topic by the Canadian Radio-television and Telecommunications Commission.

Since I have accepted working in a system where access to information and privacy protection are combined in the same act, what more could I try to do in the area of information policy? The real burdens of trying to do the former relatively well speak volumes for a policy of non-expansionism for officials with responsibilities like mine.

Complaints, Audits, and Access Rights

The section on this subject in my book presented some conclusions about the conduct of the actual core activities of data-protection agencies. My skepticism about the centrality of complaints as a guide to implementation has been borne out in British Columbia. Complaints are indeed a "safety valve" for an aggrieved public (usually the person who thinks he or she has been negatively affected) and do help to set priorities for audits or inspections by my office.[35] The latter point is especially relevant for complaints from interest groups such as the BC Civil Liberties Association. Its complaint about drug testing by the BC Racing Commission gave me an opportunity to visit a harness track and to prepare a report on drug testing in this tiny part of the public sector.

It is easy for persons who have been subject to government action, such as removal from office, to complain to my office that their privacy has been invaded, since information about them was allegedly used in an unauthorized fashion. In such a case, my portfolio officers investigate and prepare a report, which is released to the individual. There are complainants who seek government benefits but do not wish the authorities to hold or use personal information about them. My sense is that my colleagues do a fair bit of explanation of the facts of life to such individuals. I had certainly not anticipated the number of difficult clients that would find their way to a "helping" organization such as my own; university students are quite a compliant lot in comparison.

I must also admit that the Freedom of Information and Protection of Privacy Act was drafted in such a way that public bodies, including law-enforcement agencies and those who can claim to be engaged in a law-enforcement matter, enjoy considerable latitude in using personal information for purposes perceived as legitimate. This is a matter that we will have to review in detail in preparing submissions for the legislative review of the act. The public service, including municipal police forces, were not guileless in the shaping of the legislation.

My sense is that complaints to my office in British Columbia have not yet produced the systemic results that I would hope for from such a mechanism for redressing grievances. We have had much more systemic success from our eight investigative reports as a result of incidents that have occurred or, in fact, complaints. Incidentally, our frequent practice in the latter regard is to rely on separate investigative reports prepared by public bodies themselves, such as the Ministry of Health or a hospital, when privacy disasters occur. This has the particular benefit of preserving our human resources for essential activities and permitting us to audit the auditors.

I have always been an admirer of the conduct of audits or inspections by the German federal and state data-protection offices and the Privacy Commissioner of Canada for the pursuit of statutory objectives: "Audits are crucial to an activist, aggressive stance; . . . it is necessary to create an atmosphere of prior restraint for prospective privacy offenders."[36] I am especially interested in on-site inspections, which I have come to describe by the more neutral term "site visits." My most significant contribution in 1992 to the shaping of the BC Freedom of Information and Protection of Privacy Act was to urge, successfully, the inclusion of authority for the commissioner to "conduct investigations and audits to ensure compliance with any provision of this Act" (Section 42(1)(a)).

My site visits are a form of consciousness raising among public bodies. I have been to many prisons, an adolescent detention center, Correctional Services, hospitals, and municipal police forces, city halls, local Human Resources offices, psychiatric and counseling offices, universities and colleges, and public health departments. My colleague who paves the way for site visits has to engage in quiet diplomacy, since the initial reaction to a proposed inspection is not always welcoming. I meet with the head of the public body or office, the persons with direct responsibility for compliance with the act, and senior management. Whenever possible, I give a talk about the act to a gathering of the staff, and I answer questions. This has been a particular characteristic of my visit to hospitals, an area that I am very concerned about in terms of promoting fair information practices. I always do a walkabout to visit representative parts of the operation. I sample printed and electronic records, review what is accessible from staff computers, and ask questions of individual staff as I encounter them, whether in personnel offices or in the workstations of head nurses. I have the authority to look at any records held by a public body. Many small problems tend to come to the surface during my visits, and these are often easily remedied. I believe that I have especially reinforced the roles of those with delegated responsibilities for implementation of the act—the agents on the spot, so to speak. My visits also have a ripple effect, by word of mouth, among similar public bodies.

The reality is, however, that I have relatively little time to spend on site visits, if I am going to keep up with my case load of decision making. The portfolio officers from my office are often in the field mediating specific cases, and I have encouraged them to contribute to the conduct and impact of site visits. They too offer advice on the spot and suggest improvements. I intend to avoid the illusion of data protection by repeating site visits to public bodies that appear to warrant them.

Despite the efforts of my learned colleagues among privacy advocates to persuade me otherwise, my direct experience has reinforced the skepticism I expressed in my book about the importance of access and correction rights.[37] I am delighted that the Ministry of Children and Family Services has given out more than a million pages of records to thousands of applicants who want to learn the histories of their lives as children in care or as adoptees. That is a direct, consequential benefit of the act with enormous significance for individual lives. I believe, intuitively, that knowledge that a person may access information about him or her recorded by a public body engenders appropriate prudence and cautiousness among those who create and compile such records in the first place. For the large ministries that collect a lot of personal information, the Portfolio Officers on my staff mediate a considerable number of individual requests for access to personal information. Thus, I have rarely had to deal directly with a denial of access to personal information. We are also very involved in ensuring access to general information held by public bodies.

Monitoring Surveillance Technology

I advocated a special role for data-protection agencies in monitoring developments in information technology at a time when personal computers were not yet ubiquitous. In my 1989 book I wrote: "For either a well-meaning or a malevolent regime, there are no technical limits to electronic surveillance and social control at the present time."[38] The situation can only be viewed as having worsened since I wrote those words, and one fears that it will continue to deteriorate despite the best efforts of data protectors. Technological imperatives are increasingly harnessed to government's goals of reducing costs, avoiding fraud, and improving efficiency.

One can today only acknowledge "the continued and voracious expansion of the public and private sectors' appetite for more and more refined and integrated personal data at the expense of personal space and individual autonomy."[39] I am more aware of this in the private sector than in the public sector. My sense is that the public sector has not been able to afford the software and data-matching resources for personal information that have been marshalled by direct marketers (especially in the United States).

We are making an effort in British Columbia to reach the "specialists in informatics" who can "mobilize technological expertise for protective purposes."[40] But I do not have such specialists on my staff, nor could I justify such expenditures.[41] I am fortunate to have on my staff a few people with systems backgrounds and considerable interest in applications and developments in technology. I rely on their antennae and networks to keep me informed on relevant issues. The government operates various fiber-optic networks across the province, which means that we need not worry too much about the security of data being transmitted across the network. But the government's recent decision to disband the BC Systems Corporation, which ran those operations, might have had detrimental effects if its security specialists, a team of approximately a dozen, were not kept intact within the Office of the Chief Information Officer.

In 1989 I raised the issue of data protectors' functioning as legitimizers of new forms of technology.[42] This issue continues to concern me. When the Ministry of Health asked for my advice about the Pharmanet prescription profile system, which it was fully determined to implement in any event, I insisted that everyone required to participate in the program (i.e., anyone wanting any prescription filled) should have the option of a password. When I finally needed to join Pharmanet myself, I learned that I was the first one to ever ask for a password at a certain busy pharmacy; the staff had to consult a manual to learn how to give me one. I fear that most participants have no idea that it is even possible to get a password with which to control access to their own prescription records.

We are having modest success in British Columbia in promoting the preparation of privacy impact assessments by public bodies introducing new forms of technology and new or significantly altered personal-information systems.[43] The Ministry of Health is a leader in this. When a proposal of this sort reaches my office with an accompanying impact statement, it is a considerable blessing from every perspective, including the protection of human rights, because the proponents have already thought about fair information practices at the design stage of whatever they are contemplating. We do not have the resources to prepare similar documentation except under duress in the course of crisis management.

Strengthening Data-Protection Legislation

My office is preoccupied with its assigned tasks every day. For this reason, opportunities to look around and remember the big picture are built into our work program. The professional members of the staff periodically go on one-day retreats. We regularly hold all-staff luncheons, at which a number of talented privacy specialists have spoken. And in 1996 we co-sponsored an international conference in Victoria on the theme of Visions of Privacy for the 21st Century.

Since the BC Freedom of Information and Protection of Privacy Act is typically general, we are encouraging public bodies to incorporate specialized fair information practices into revised legislation or regulations dealing with specialized activities.[44] I believe strongly that general principles should be incorporated in sectoral legislation, over time, as it is revised, so long as the privacy commissioner continues to have oversight of its detailed functioning.

A related matter is our ongoing effort to require privacy codes to be in place for those who collect or receive personal information on behalf of public bodies. Thus I am encouraging the Insurance Corporation of British Columbia to require a privacy code for its Autoplan agents, who insure all motor vehicles in the province. I am also concerned with ensuring that the thousands of service providers and contractors for the Ministry of Health and the Ministry of Children and Family Services implement fair information practices as required by the Freedom of Information and Protection of Privacy Act. The 1996 Model Privacy Code of the Canadian Standards Association is an ideal vehicle for additional self-regulation.

What remains problematic in British Columbia, despite the European Union's 1995 Directive on Data Protection, is an effort to extend data protection to the private sector. I still optimistically believe that the situation will change in the next several years, but I have to admit that my written and oral efforts to stimulate discussion of the matter have fallen on deaf ears to date.[45] The fact that Quebec became the first jurisdiction in North America to so legislate is, of course, encouraging. I wrote in 1989 that "statutory data protection is also essential for the private sector," and that "the long-term goal must be to ensure individual rights in all spheres of human existence." It was not practical to lobby the BC government on this score while the Freedom of Information and Protection of Privacy Act was being implemented in three tiers over three years; however, the reelection of the New Democratic government in 1996 presents new opportunities.

The Freedom of Information and Protection of Privacy Act is quite progressive, even by Canadian standards. It now extends to 33 self-governing bodies of professions or occupations, including the College of Physicians and Surgeons and the Law Society of British Columbia. I hope to encourage self-governing professional bodies to promote self-regulation among their members via privacy codes. I am also encouraging a review of the adequacy for the next century of the provisions respecting privacy in the BC Credit Reporting Act.

Toward the Future

In 1989 I boldly asked what data-protection authorities would look like by the year 2000, which seemed far enough in the future that I would not have to face the consequences of what I wrote. I did not anticipate that I too would be in part responsible for that future vision, since my term ends in 1999. I raised the prospect that we would be perceived as "a rather quaint, failed effort to cope with an overpowering technological tide," and I said it was self-evident that "data protection agencies will have to be vigilant, articulate, and resourceful in fashioning acceptable solutions in the public interest."[46] Although I believe that my office has been vigilant and resourceful to date, I am less inclined now to draw an optimistic conclusion than I was in 1989, because of the ongoing explosion of the digital economy and online Internet services.

Acknowledgements

I am grateful to the following colleagues: P. E. (Pam) Smith, Lorrainne Dixon, and Kyle Friesen.

Notes

1. David H. Flaherty, *Protecting Privacy in Surveillance Societies* (University of North Carolina Press, 1989).
2. Ibid., p. 377.
3. Ibid., p. 373.
4. Ibid. , p. 375.

5. Ibid., p. 377.

6. See William A. Parent, Privacy: A brief survey of the conceptual landscape, *Santa Clara Law and Technology Journal* 11 (1995), no. 1: 21–26.

7. *Protecting Privacy in Surveillance Societies*, p. 8.

8. Ibid., p. 380.

9. Ibid., p. 379.

10. Ibid.

11. Ibid., p. 380.

12. Ibid., p. 406.

13. Ibid., p. 383.

14. Ibid., p. 383.

15. Ibid., p. 383.

16. Ibid., p. 384.

17. Ibid., p. 384.

18. Ibid.

19. Ibid., p. 385.

20. Ibid., p. 386.

21. Ibid., p. 387.

22. Ibid., p. 387.

23. Ibid., pp. 388–389.

24. Ibid., p. 389.

25. Ibid., p. 391.

26. Ibid., pp. 41–44, 113, 260, 414.

27. Ibid., p. 393.

28. Ibid., p. 393.

29. Ibid., p. 394.

30. Ibid., pp. 394–395.

31. Ibid., p. 395.

32. Ibid., p. 396.

33. Ibid., p. 397.

34. Ibid., pp. 397–398.

35. Ibid., p. 400.

36. Ibid.

37. Ibid., pp. 401–402

38. Ibid., p. 402.

39. Ibid., p. 403.

40. Ibid.
41. See ibid., p. 404.
42. Ibid., pp. 403–404.
43. Ibid., p. 405.
44. Ibid., pp. 404–405.
45. Ibid., p. 406.
46. Ibid., pp. 406–407.

7

Does Privacy Law Work?

Robert Gellman

A logical first step in an evaluation of the law as a mechanism for regulating privacy is to define the universe of privacy law. This may be an impossible task. To identify privacy laws, we must decide what is meant by privacy. In the United States, privacy can be a broad and almost limitless issue. Privacy is cited to include everything from control over personal information to personal reproductive rights to limits on government intrusion into the home. The US Constitution guarantees individuals a right to be secure in their persons, houses, papers, and effects against unreasonable searches and seizures; a right against self-incrimination; and a right to speak and assemble. These and other constitutional principles are routinely associated with privacy interests, but the word "privacy" appears nowhere in the Constitution.

Lawyers, judges, philosophers, and scholars have attempted to define the scope and the meaning of privacy, and it would be unfair to suggest that they have failed. It would be kinder to say that they have all produced different answers. One of the most quoted formulations comes from Louis Brandeis, who consistently referred to privacy as "the right to be let alone."[1] Some question whether this is a meaningful consideration in today's world, in view of the abundance of third-party record keepers and the near-impossibility of keeping personal information from being maintained in other people's computers.[2] Other schools of thought define privacy in a broader social context, with more of a focus outward toward society rather than inward toward the individual.[3] Because a good deal of the development and nurturing of the legal right to privacy has taken place in law journals, the confusion and lack of agreement about the concept may be entirely understandable.

The problem of defining boundaries is not a trivial one. Under any definition, everyone is likely to agree that privacy interests are affected by statutes such as the Fair Credit Reporting Act,[4] Family Education Rights and Privacy Act of 1974,[5] and the Electronic Communications Privacy Act,[6] and by constitutional provisions such as the Fourth Amendment's guarantee against unreasonable searches and seizures.

The lack of a clear definition is more critical at the margins. For example, the Intelligence Identities Protection Act of 1982[7] makes it a crime to identify and expose covert intelligence agents. This law protects information about individuals, but it has a sharply different history and motivation than the traditional privacy laws cited above. Its purpose was to punish those who publicly identify secret government agents. Many other laws address, in some manner, the collection, use, and disclosure of personal information, or use the word "privacy." Drawing a line is difficult.

There is no reason to believe that a new effort to define privacy or even to establish clear categories for privacy law will achieve consensus. In 1995 Alan Westin, a long-time privacy scholar, said at a conference that no definition of privacy is possible, because privacy issues are fundamentally matters of values, interests, and power.[8] In any event, given the task at hand—to consider the role of law in regulating privacy—it is possible to proceed without defining terms more precisely. Those who are responsible for the law—the courts and the legislatures—do not necessarily operate in logical, consistent, and organized ways. We can make some progress here without constructing a comprehensive framework for evaluating their activities.

However, to establish some focus, the attention here will be mostly, but not exclusively, on the slice of privacy known as "data protection." This is a useful European term referring to rules about the collection, use, and dissemination of personal information. One major policy objective of data protection is the application of fair information practices, an organized set of values and standards about personal information defining the rights of record subjects and the responsibilities of record keepers. This is an important subset of privacy law. Privacy law and policy throughout the world have converged around fair information practices.[9] The goals of most modern privacy or data-protection laws can be described with the same set of fair information principles.

Whereas in Europe data-protection laws tend to establish omnibus rules that apply to most records (public and private), in the United States few laws regulate privately maintained records in the interest of privacy. The American approach to privacy is sometimes called "sectoral." There are no general privacy laws, just specific laws covering specific records or record keepers. As a result, the legal structure for privacy in the United States is a patchwork quilt. This makes for more interesting analysis, although not necessarily for better privacy protection. Out of the disorganized welter of legal activity relating to privacy in the courts and the Congress, three main sources of privacy law can be found: constitutional protections, common-law remedies, and statutes that attempt to address privacy concerns. What kind of privacy protections can we expect from each of them? Can these sources of law effectively address the objectives of fair information practices?

Statutory Codes of Fair Information Practices

Concerns about the power of government to use and abuse personal information—especially in a computerized environment—produced their first federal legislative response: the Privacy Act of 1974.[10] This law establishes general rules for the collection, maintenance, use, and disclosure of personal information held by federal agencies. The Privacy Act is worth reviewing here to describe the code of fair information practices and to evaluate the law.

A key objective of the Privacy Act was restricting the government's use of computer technology to invade privacy. This act was based on the 1973 recommendations of a federal advisory committee.[11] The Secretary of Health, Education and Welfare convened the committee in response to the growing public and private use of automated data systems containing information about individuals. The congressional hearings that led to the Privacy Act also focused on the government's use of computers.[12] The act included an express finding that computers "greatly magnified the harm to individual privacy that can occur from any collection, maintenance, use, or dissemination of personal information."[13] Thus, the executive and legislative branches looked at the increasing computerization of personal records and decided that new controls on technology were needed and that new protections for individuals were appropriate. The Privacy Act set forth clear goals. The question is whether the act is effective in meeting those goals in the face of growing computerization.

In some ways, the Privacy Act was a tremendously influential piece of legislation. It was the world's first attempt to apply the principles of fair information practices. The notion of fair information practices came directly from the work of the HEW advisory committee. The articulation of principles of fair information practices may be the computer age's most significant policy development with respect to privacy.

This does not mean that the Privacy Act was a success at home. There is a big difference between adopting good policies and implementing them well. A review of the act under the framework of fair information practices illustrates the statutory and administrative shortcomings.

The first principle of fair information practices is openness or transparency: there should be no secret record-keeping systems. The Privacy Act succeeded (at least initially) in meeting this objective by requiring that descriptions of record systems containing personal information be published. Before the various agencies prepared this inventory, few had any idea what data they maintained. Openness was a positive exercise in good administration and good records management. There were redundant and unnecessary systems throughout the government, and many were eliminated.

Between 1975 and 1990, the number of personal-record systems shrank from almost 7000 to under 5000. On the surface this seems an improvement, and it may be. It is difficult to determine whether the reduction really increases privacy protection. Some reductions resulted from mergers of existing systems. When different record systems with different users are merged, the protections that followed from separation may disappear. The result may be that more government employees have access to disparate information in common computer systems. This does not necessarily enhance privacy. Fears about a single computerized government data file on individuals contributed to passage of the Privacy Act in the first place.

The initial burst of file cleaning that took place when the Privacy Act took effect in 1975 has long since dissipated. Strong evidence suggests that agencies grew sloppy and lost interest in keeping up with their obligations to publish accurate descriptions of their record systems.[14] The absence of an effective oversight mechanism to keep pressure on agencies may be one reason for this.

The second principle of fair information practices is individual participation. The subject of a record should be able to see and correct the record. The Privacy Act's procedures for access and correction appear to work adequately for US citizens and resident aliens, who have rights under the Privacy Act. Foreigners have no access rights or correction rights.[15]

There has been considerable litigation seeking enforcement of correction rights. The Privacy Act created an outlet for the tension between the record-keeping practices of the government and the interests of identified individuals. Because privacy laws are not always self-implementing or self-enforcing, they tend to be more successful when specific individuals have an incentive to pursue their own interests. This tension is missing for many other parts of the Privacy Act.

The third principle of fair information practices is that there should be limits on the collection of personal information. The fourth principle is that personal data should be relevant to the purposes for which they are to be used and should be accurate, complete, and timely. The Privacy Act's implementation of these two principles is just as vague and general as the principles themselves, and it is difficult to measure the actual impact of the act. Anecdotal evidence suggests that agencies rely on these requirements mostly when they provide excuses for avoiding an unwanted task.

There is, however, no specific enforcement or oversight of these information-collection principles. In sharp contrast, the Paperwork Reduction Act requires an elaborate administrative approval process before a federal agency collects information.[16] The purpose of the paperwork law is to reduce the burden of the government's information demands on individuals and businesses. Congress's willingness to establish a bureaucratic process to reduce paperwork but not to enforce the goals of privacy laws speaks for itself. The agency that oversees the Paperwork Reduction Act is the Office of Management and Budget, the same office that is responsible for assisting agencies with the Privacy Act.[17] However, the OMB devotes far fewer resources to the Privacy Act than to the Paperwork Act.

The fifth principle of fair information practices requires limits on the internal use of personal data by the record keeper. This is where the major shortcomings of the Privacy Act begin to become clearer. The act limits use of personal data to those officers and employees of the agency maintaining the data who have a need for the data in the performance their duties. This vague standard is not a significant barrier to the sharing of personal information within agencies. The need-to-know standard is not related to the purpose for which the records were collected. The law provides no independent check on self-serving assessments by agency officials of the need to know. No administrative process exists to control or limit internal agency uses. Suits have been brought by individuals who objected to specific uses, but most uses have been upheld.[18]

The sixth principle prohibits external disclosure without the consent of the subject or other legal authority. This is almost certainly the biggest failure of the Privacy Act. A standard rendition of the policy goal is that records should be used or disclosed only for the purpose for which they were collected. Accomplishing this objective when the purposes are not clearly defined and are subject to constant change is a challenge. The authors of the act recognized the impossibility of establishing a fixed statutory test for judging external disclosures of personal information from the myriad of record systems maintained by federal agencies. The legislation left most decisions about external uses to the agencies, and this created the biggest loophole in the law.

An agency can establish a "routine use" if it determines that a disclosure is compatible with the purpose for which the record was collected.[19] This vague formula has not created much of a substantive barrier to external disclosure of personal information. Agencies generally operate under the belief that they can disclose any record for almost any purpose if the law's procedural requirements of publishing a notice in the *Federal Register* is met. Limited oversight of routine uses by the OMB and by the Congress had little general effect.[20] Later legislation, political pressures, and bureaucratic convenience tended to overwhelm the law's weak limitations. Without any effective restriction on disclosure, the Privacy Act lost much of its vitality and became more procedural and more symbolic.

The weakness of the Privacy Act's disclosure policies became fully apparent with the development of computer matching in the late 1970s as a technique for overseeing government benefit programs. Matching identifies individuals enrolled in more than one benefit program by comparing computer tapes. The presumption behind matching—not borne out at times—is that anyone enrolled in two overlapping programs receives some benefits improperly.

Computer matching offered an important test for the Privacy Act. Would the Privacy Act restrain new uses of computer technology that cut across existing administrative and functional safeguards? Initially, the routine-use provision of the Privacy Act imposed a partial barrier to computer matching. Investigating food stamp fraud is not related to the purpose for which federal personnel records are collected, and disclosure of personnel files to fraud investigators is arguably incompatible with that purpose. However, the political pressures to allow data matching were enormous, and there were sharp attacks on the initial interpretation of the act that limited the use of computer matching. The Office of Management and Budget, which has some oversight responsibilities for the Privacy Act, was not willing to question or interfere in any significant way with computer matching. The OMB issued mildly procedural rules for matching in 1979 and weakened them in 1982. In any event, federal agencies mostly ignored the guidance, and the OMB did not enforce it.[21] Congress eventually enacted new laws directing agencies to engage in computer matching for specific programs, and this removed some but not all of the questions about the legality of matching.

Despite its origins in congressional and executive concerns about the effect of computerization on privacy, the Privacy Act failed to prevent or control a significant expansion of computerized manipulation of personal records through matching. One reason for this is that computer matching is a politically attractive technique that aims at large classes of individuals. It focuses on identified individuals only at the end of the process, when individual wrongdoers have been identified (correctly or incorrectly). There is no tension at the "front end," which serves as a barrier to matching and to the interconnection of previously separate record systems.

In an attempt to regulate the use of computer matching, Congress passed the Computer Matching and Privacy Protection Act of 1988.[22] That law did not establish substantive restrictions on matching activities. It only erected procedural hurdles to the approval of specific computer matching programs, including better notice, oversight by agencies' data-integrity boards, cost-benefit requirements, and mandatory due process.

However, a review of the new procedures by the General Accounting Office in 1993 found that the law had produced limited changes in the oversight of computer matching. The GAO found that agencies did not fully and earnestly review proposed computer matches and that required cost-benefit analyses tended to be done poorly or not at all.[23] The bureaucracy largely absorbed or ignored the procedural impediments to matching programs. The newly created data-integrity boards did not operate with real independence. Anecdotal evidence suggests that the new bureaucratic process deterred some proposals for data matching. Dedicated privacy officers in a few agencies used the procedures to stifle inappropriate matching. For the most part, however, the Computer Matching and Privacy Protection Act resulted in few real restraints on data matching.

The seventh principle of fair information practices calls for a reasonable level of security. The Privacy Act includes a general requirement for appropriate administrative, technical, and physical safeguards to protect information against anticipated threats.[24] It is difficult to assess general compliance with this statutory standard. The adequacy of computer security for classified national-security information and for government computers in general has been repeatedly called into question, and there are almost certainly equivalent deficiencies for Privacy Act security. There is no effective oversight of the security requirement.

The final principle of fair information practices requires that record keepers be accountable for compliance with the other principles. The Privacy Act contains civil and criminal penalties for violations, but it is far from clear that the enforcement methods are useful. When the act became effective, federal employees viewed criminal prosecution as a real threat, and they exercised a good deal of caution. This caution disappeared rapidly as people became familiar with the law and learned that actual enforcement was unlikely. In more than 20 years, federal prosecutors have brought no more than a handful of criminal cases, and perhaps only one, under the Privacy Act.

The basic method for enforcing the Privacy Act is the individual lawsuit. Aggrieved individuals can sue the government for violations. There has been a considerable amount of litigation, but a distressingly large number of cases have been brought by disgruntled federal employees seeking to stage collateral attacks on unfavorable personnel actions. The former General Counsel to the Privacy Protection Study Commission testified that the act was "to a large extent, unenforceable by individuals."[25] The main reasons are that it is difficult to recover damages and that limited injunctive relief is available under the law. Individual enforcement does not offer any significant incentive for agencies to comply more carefully with the Privacy Act's provisions. Still, some case law has cautioned agencies about loose interpretations of the act.

It would be too glib to call the Privacy Act a failure. The act addresses the full range of fair information practices, and it applies them to all government records about individuals. That is a significant challenge, and the act has had some successes. Yet the act clearly failed to offer any real resistance to broader and unforeseen uses of personal information supported by computers. This failure is due in part to the vagueness of the act and in part to the lack of oversight and enforcement. The code of fair information practices has not become ingrained into agency policies and practices. Compliance with the Privacy Act is an afterthought at most agencies.

The experience with the Privacy Act can be usefully compared with the experience with the Freedom of Information Act (FOIA),[26] which established a procedure under which individuals may request access to any records held by the federal government. Only a fraction as much litigation resulted from the Privacy Act as from the FOIA.[27] Litigation has been crucial to the development and enforcement of the FOIA. Especially after the 1974 amendments added teeth to the FOIA, vigorous requesters and sympathetic courts constantly pressured agencies to apply the law fairly. The Privacy Act is rarely pursued with the same enthusiasm or effect by litigants. Key decisions under the Privacy Act, such as the creation of exemptions and the promulgation of routine uses, go unchallenged because there are no incentives for individuals or groups to object to these actions when they are initially proposed. The limited administrative and congressional oversight is a poor substitute. The result is a law with limited effect on the bureaucracy and with no organized constituency to press for change.

 Other privacy statutes define the responsibilities of record keepers and the rights of consumers. The most important privacy law regulating private-sector use of personal information is the Fair Credit Reporting Act (FCRA).[28] The FCRA was the first privacy law of the modern era, dating back to 1970. While the law predates the code of fair information practices, the specific statutory provisions match up well with the code. The FCRA includes access and correction rules; limits on collection, use and disclosure; remedies at law; and administrative enforcement. This is not to suggest that the law necessarily works well in its current form. Industry and public-interest groups agree that the FCRA is out of date. Ed Mierzwinski of the US Public Interest Research Group described it as "a piece of prehistoric junk."[29] Efforts to revise the FCRA that had been underway since the 102nd Congress finally succeeded in 1996.

There are also federal privacy laws for video-rental records,[30] cable-viewing records,[31] and educational records.[32] Scattered other laws address limited privacy concerns. For example, the Right to Financial Privacy Act[33] requires that customers be notified of government subpoenas for bank records in some circumstances but does not address most elements of the code of fair information practices. The Telephone Consumer Protection Act[34] requires the Federal Communications Commission to issue regulations to allow residential telephone subscribers to avoid receiving telephone solicitations to which they object. Except for the FCRA, however, there is no effective administrative oversight of these laws.

Constitutional Protections

The Supreme Court has addressed privacy issues in many different types of opinions, but the discussions of information privacy are especially unclear and unsatisfying. The pronouncements most often cited reflect the lack of focus and clear boundaries for any aspect of privacy. In *Griswold v. Connecticut*, a case involving the constitutionality of a state statute restricting the availability of contraceptives, Justice William O. Douglas described privacy as found within the "penumbra" of the Bill of Rights.[35] This is a wonderful sentiment, but clearly defined protections for privacy interests do not emerge from the shadows. Justice Brandeis repeated his description of privacy as the right to be let alone in his famous dissent in *Olmstead v. United States*,[36] but the statement does little to explicate what that right means.

Olmstead is an important case in the constitutional history of privacy. In a 5-4 decision issued in 1928, the Court ruled that neither the Fourth Amendment prohibition against unreasonable searches and seizures nor the Fifth Amendment protection against self-incrimination provided protection against wiretapping. The Court found no physical trespass and no search or seizure. Brandeis dissented, arguing that the Fourth Amendment protected individual privacy and warning against "the progress of science in furnishing the Government with means of espionage."[37] This comment echoed sentiments from the famous 1890 *Harvard Law Review* article by Louis Brandeis and Samuel Warren on the right to privacy.[38] In many ways, the battle over privacy in the twentieth century has been a struggle over adapting privacy principles to constant technological advances. This conflict was presaged in that article. New technology was the principal concern that produced the call for more formal legal responses to privacy invasions. The development of "instantaneous photographs"—pictures that could be taken surreptitiously and without formal sittings—sparked new privacy theories.

It was not until 1967 that the Supreme Court overturned *Olmstead*. In *Katz v. United States*,[39] the Court ruled that wiretapping constituted a search under the Fourth Amendment and that physical trespass was not required because the Fourth Amendment protects people and not places. In a concurring opinion, Justice John Harlan set out what has become a highly influential two-part formula for assessing whether governmental action violates the Fourth Amendment. The first question is whether a person has exhibited an actual or subjective expectation of privacy. The second question is whether that expectation is one that society deems to be reasonable.

The majority opinion in *Katz* made it clear, however, that the Fourth Amendment does not provide a general constitutional right to privacy:

That Amendment protects individual privacy against certain kinds of governmental intrusion, but its protections go further, and often have nothing to do with privacy at all. Other provisions of the Constitution protect personal privacy from other forms of governmental invasion. But the protection of a person's general right to privacy—his right to be let alone by other people—is, like the protection of his property and of his very life, left largely to the law of the individual States.[40]

The wiretapping cases, then, are mostly about limiting the intrusive powers of government. This is an important privacy theme with a strong and well-developed constitutional basis. It has proved more difficult, however, to extend constitutional protection from physical intrusions to what might be called informational intrusions. This has been left, as the Court suggested in *Katz*, to statutes.

In 1977 the Supreme Court directly raised the possibility of a constitutional right of informational privacy in *Whalen v. Roe*.[41] The case involved a privacy-based constitutional challenge to a New York State statutory requirement that the names and addresses of all persons who had obtained certain prescription drugs be reported to the state and stored in a central computerized databank. The Court did not find a constitutional violation, but the opinion did include another tantalizing comment about constitutional privacy rights. The Court stated that the duty to avoid unwarranted disclosures "arguably has its roots in the Constitution."[42] This statement hints at the existence of a constitutional right of informational privacy, but the Court did not squarely hold that the right exists. The Court observed that disclosures of private medical information are often an essential part of modern medical practice. The Court could not conclude that reporting to the state was an impermissible invasion of privacy.

The opinion reflects the theme of privacy and technology. The Court expressed concern over the "vast amounts of personal information in computerized data banks or other massive government files." The decision to uphold the statute was strongly influenced by specific protections against unauthorized use and disclosure in the computerized reporting system. Had there been a greater possibility of public disclosure under the New York statute, the Court might have confronted the constitutional interest more directly. Where this might have led is uncertain.

In the end, the Court seemed to say that the New York statute did not pose a sufficiently grievous threat to a constitutional interest, no matter how that interest might be defined. In a concurring opinion, Justice William Brennan said that he would not rule out the possibility that there may, at some point, be a need for a curb on technology that supports central storage and easy accessibility of computerized data. In a separate concurrence, Justice Potter Stewart expressed his view that there is no general constitutional right of privacy, although there are protections against government intrusion into personal and private matters. These two concurrences illustrate the hope and the reality of constitutional protections for information privacy. Not surprisingly, *Whalen* created a considerable amount of confusion in the lower courts, and cases can now be found on all sides of the constitutional right of information privacy.[43]

A case decided by the Supreme Court a year before *Whalen* illustrates the lack of any constitutional protection for personal information maintained by third-party record keepers. In *United States v. Miller*,[44] the Supreme Court decided that an individual had no expectation of privacy in account records held by a bank. The account owner was not entitled to notice or the opportunity to object when the government sought records of the account from the bank. In view of the extensive amount of personal information (financial, medical, transactional, etc.) maintained routinely in records of third-party record keepers, the *Miller* decision shows that an individual has no privacy interest in those records even when the government is seeking access.

The absence of a clear constitutional right of informational privacy is further illustrated by a successful challenge to the disclosure of Social Security numbers from Virginia's voter-registration records in *Greidinger v. Davis*.[45] The plaintiff challenged the public-disclosure requirement of the law as an unconstitutional burden on the right to vote.[46] No constitutional right of privacy in the Social Security number was asserted. The plaintiff argued only that the privacy interest in the number is sufficiently strong that the right to vote cannot be predicated on disclosure of the number to the public or to political entities. The Fourth Circuit Court of Appeals agreed, quoting *Whalen* to make a point about the dangers of computerized data banks, but the case turned entirely on the constitutional interest in voting and not any constitutional interest in informational privacy.

In *Whalen*, the Supreme Court applied—albeit without acknowledgment—security policies that are part of fair information practices. Policy and even statutory statements about security are typically vague. Absent clearer direction or standards, the Court was unable and unwilling to overturn a law that included at least some protections against unauthorized disclosure and use. Perhaps a poorly written law or a better factual showing of actual privacy violations will prompt a more aggressive opinion someday. It is helpful that the Court was willing to consider the security aspect of fair information practices in its constitutional analysis, even if the decision did not go further than it did. However, there is no real suggestion that fair information practices are a constitutional requirement.

It is difficult to see how constitutional litigation would ever address fair information standards such as openness, access, and even accountability. As *Katz* shows, the Constitution clearly offers strong protections for some privacy interests against some forms of governmental intrusion. *Katz* did not define the specific procedural steps appropriate for regulating wiretapping. This was appropriately left to the Congress. *Greidinger* is less helpful; it was able to address privacy only in the context of protecting a well-established constitutional right, namely the right to vote. *Miller* shows that the constitution offers no protection against access to third-party records.

Of course, the US Constitution will never restrict the use of personal information by private parties. Even with governmental information practices, however, the code of fair information practices cannot be mapped onto existing constitutional privacy provisions. This is not a surprising conclusion, but it emphasizes that solutions for many current privacy concerns will not be found in the Constitution. There is no firm basis to be found in Supreme Court decisions for asserting any general notion of a constitutional right to informational privacy. The same concerns reflected in fair information practices may influence holdings in some constitutional cases, as happened in *Whalen*. Still, the Constitution is not a device for imposing most fair information practices on government. For substantive and procedural privacy protections, especially for personal information, we must look elsewhere.

Before moving on, however, it is worthwhile to compare the statutory aftermath of the *Katz* decision with the Privacy Act of 1974. Both laws regulate the federal government. We have already found the Privacy Act to be a less than fully effective statute. The wiretap law is more successful.

Title III of the Omnibus Crime Control and Safe Streets Act of 1968 quickly followed the *Katz* decision. Congress took up the suggestion of the Supreme Court and defined the conditions and circumstances under which the government could intercept wire and oral communications.

The law represented a balance between privacy interests and the needs of law enforcement. The statutory requirements for wiretapping are strict. A high-ranking government official must approve the request for a court order. The government must prove that probable cause exists for believing that a crime had been committed, that the target of the surveillance is involved, and that evidence would be obtained though the surveillance. The law also includes requirements for minimization of surveillance.

The legislative battles that ultimately produced Title III lasted 40 years. Initial criticism of *Olmstead* was widespread, and members of Congress regularly introduced bills to allow or prohibit wiretapping. Support for wiretapping waxed and waned as the nation moved through Prohibition, World War II, the McCarthy period, the Cold War, and other events. For 40 years there was no lack of congressional attention, interest, or understanding of the issues involved in wiretapping. What was lacking was consensus. Congress simply could not reach agreement on a suitable policy that would either regulate or prohibit wiretapping.

It was not until the Supreme Court's 1967 *Katz* decision that the burden of going forward with legislation shifted. When wiretapping was constitutional, its proponents had little incentive to agree to restrictive legislation. The status quo benefitted wiretappers. Once wiretapping without warrants was held to be unconstitutional, both proponents and opponents had something to gain from legislation. The sharply held views on both sides did not change, but the legislative dynamics did. After decades of legislative stalemate, the *Katz* ruling quickly prompted a compromise.

The wiretapping law has been generally successful in accomplishing its objectives. It is not strictly a fair-information-practices law, although it does contain similar elements. Why has Title III worked for the most part while the Privacy Act has a much spottier record?

First, in debates over wiretapping debates, it was easy to explain the consequences of unrestricted government wiretapping and the need to limit government eavesdropping on telephone calls or electronic messages. People understand what it means when a government agent is listening to their telephone calls and that wiretapping, under anybody's rules, is an extraordinary event. The goal of the wiretapping law is more clearly and more narrowly defined. In contrast, the Privacy Act has many goals, not all of them precise. Controlling the government's use of personal information through fair information practices is a more abstract objective. The public generally accepts that the government legitimately maintains records about individuals in order to carry out its activities. People also eventually became comfortable with the use of computers to manipulate that information. Even privacy professionals sometimes find it difficult to draw a clear line between proper and improper uses of personal information maintained by the government.

Second, communications privacy benefitted from a preexisting framework for analysis. Once the Supreme Court determined that wiretapping was a search under the Fourth Amendment, familiar rules and procedures were available to regulate and evaluate conduct. The challenge was to enact the right balance of procedural protections from a toolbag of customary choices. In contrast, when the Privacy Act was passed, the code of fair information practices was a new concept. The code became highly influential among policy makers, but it did not and has not become a popularly understood concept. There is no lengthy history of judicial interpretation as there is with the Fourth Amendment. There are no constitutional absolutes to guide interpretation and understanding. Without an external frame of reference or significant oversight, the bureaucracy was able to interpret the Privacy Act to suit its own convenience.

Third, the wiretapping laws regulate the activities of a small group of government officials. Even within this small group, wiretapping is an infrequent activity. With a narrow focus, it is easier to control behavior. In contrast, the Privacy Act imposes requirements on every government officer and employee. It covers thousands of record systems and billions of records. Regulating widespread behavior in a large bureaucratic enterprise is obviously a greater challenge.

Fourth, a law-enforcement agency seeking a wiretap must obtain a search warrant from a judge. Independent oversight of individual wiretap requests is built into the process. The wiretap law establishes a framework for weighing individual privacy interests against particular law-enforcement needs. In addition, many individuals aggrieved by wiretapping end up in court in criminal matters and have the ability and the incentive to raise objections about violations of the wiretap rules. Active oversight of wiretapping comes at both ends of the approval process.

In contrast, the Privacy Act establishes broad rules for the treatment of classes of records. Agencies resolve most issues on their own, without any independent oversight and typically without a focus on specific individuals. The Office of Management and Budget has responsibility to issue guidelines and regulations and to provide assistance to and oversight of agencies' Privacy Act activities.[47] Over the years, however, the OMB has done little oversight of agencies' Privacy Act operations. Agencies largely have a free hand to interpret the act as they please. When provisions are ignored, there is often no one to object. Most other countries that have privacy laws have established specialized agencies to oversee and enforce those laws. The privacy scholar David Flaherty suggests one reason American privacy laws are not effective: "The United States carries out data protection differently than other countries, and on the whole does it less well, because of the lack of an oversight agency."[48]

The clear conclusion is that a privacy law is more likely to be successful when there is independent oversight and enforcement. This is the element largely missing from the Privacy Act. Litigation has been only partly successful under the Privacy Act. There are no constitutional principles that serve to give the judiciary a direct role in the protection of informational privacy.

Common-Law and Other Forms of Privacy Protection

Limiting the power of government is a well-established doctrine of American political philosophy, and it poses a distinct set of questions. For most of American history, government was perceived as the principal threat to personal privacy interests. Controls on the government's ability to arrest or investigate individuals, to regulate behavior, and to collect and use information were high priorities. The importance of such controls has not diminished. Privacy threats from government have now been joined by threats from the private sector. Records maintained by private institutions—employers, physicians, insurers, credit reporting agencies, banks, credit grantors—play an increasingly important role in the lives of individuals.

In the twentieth century, tort law developed some common-law remedies for invasions of privacy. These remedies are largely but not exclusively aimed at private rather than governmental actions. Four distinct privacy torts are commonly recognized: intrusion upon an individual's seclusion or solitude, public disclosure of private facts, placing an individual in a false light highly offensive to a reasonable person, and unpermitted use for private commercial gain of a person's identity. In addition, the right of publicity—or the right to control commercial use of an individual's identity—is often recognized as a related right. In 1960 William Prosser described the four basic privacy torts in a classic law-journal article.[49]

The question, however, is whether these remedies are relevant and responsive to the privacy concerns of the computer era. Can privacy tort law provide an individual with an effective response to the ongoing creation of private databanks with individual profiles? Some information comes from private sources, such as credit reports and transaction records. Some comes from public sources, such as drivers' licenses, motor vehicle registrations, occupation licenses, voter registration, land titles, hunting licenses, and court records. Few statutes regulate use of this information in private hands. Companies consult private databases containing consumer profiles to decide if individuals should be employed or should be allowed to purchase insurance, have credit, rent an apartment, open a bank or brokerage account, see a doctor, or engage in other common and basic human activities.

Can tort law really provide meaningful remedies for individuals? Litigation begun in Virginia in 1995 questions the sale of individual names through the rental of mailing lists, relying on the tort of misappropriation of name or likeness. The plaintiff lost, and an appeal is pending. The ultimate resolution of this lawsuit may give a better idea whether tort law will be a tool for aggrieved consumers against the common practice of buying and selling consumer information. A ruling that the sale of a mailing list violates the rights of the individuals on the list would have a significant effect on the marketing and mailing-list businesses.

No matter how big a victory might be obtained through a tort action, the scope of any relief will not meet the broad objectives of the code of fair information practices. Most elements of the code are not attainable through tort litigation. For example, the classic privacy torts are not likely to induce or force a record keeper to publish descriptions of record systems, to limit collection practices, to meet data quality standards, to allow individual access and correction, or to restrict internal uses of data.[50] Restrictions on the disclosure of personal data may be a possible remedy for the tort of appropriation of name or likeness. Even here, black-letter privacy tort law provides that there can be no liability for use of public-record information,[51] and considerable personal information with commercial value is publicly available in government files. It would take a major expansion of existing remedies to limit the use of public-record information. Since common-law tort remedies are statutory in many states, creative extensions by the courts are unlikely.

Tort remedies may respond to some privacy concerns, but they do not match up with the reality of modern information technology. Because of the generally hidden nature of the commercial exchange, compilation, and use of personal information, existing tort law does not reach far. For the most part, personal information is used inside organizations in ways that are not visible to the outside world or to the subject of the information. There is no physical intrusion or public disclosure. No false light is shed. Whether there is an appropriation for an unpermitted use is more of an open question. And there is a second shortcoming. Paul Schwartz speaks about the "silent ability of technology to erode our expectations of privacy."[52] When privacy standards are defined in terms of expectations or by a "reasonable person" test, the fight may already be over. The widespread use of computers to collect, combine, and manipulate personal information may have already redefined the standards. Privacy interests lose without a struggle because technology comes without any inherent privacy restrictions. Once the use and the manipulation of data have become commonplace and profitable, opponents are hard pressed to argue successfully that those activities are unreasonable. Perhaps the best argument will be that consumers are unaware of the capabilities of technology and therefore have no contrary expectations.

If restrictions on the use of personal information are to come through tort remedies, new approaches will be needed. That the common law will be able to respond in any effective way is unlikely. The lengthy and complex requirements of the code of fair information practices beg for statutory definition. Yet, as we have seen, there is no guarantee that a fair-information-practices statute will be effective.

In theory, contractual agreements between record keepers and record subjects may provide some useful remedies. This has not always been the case to date. For medical records, where there has traditionally been a strong interest in confidentiality and a long-standing ethical obligation by physicians to protect the confidentiality of patients, litigation is not an especially fruitful remedy. According to the 1977 report of the Privacy Protection Study Commission, a patient who sues a doctor for unauthorized disclosure is likely to lose.[53] A suit against one of the many new institutions (such as claims processors and outcomes researchers) that routinely maintain identifiable health information would be even harder to win.

In other circumstances where a record keeper's obligation is less clear or where the terms of a contract are determined largely by the record keeper without any active bargaining with the record subject, contractual remedies are not likely to help record subjects. Nevertheless, this may be an area where additional developments are possible. In environments where privacy is a significant concern of consumers, as with electronic mail and online services, vendors may have incentives to offer stronger assurances of privacy to customers. If merchants detect that a sizable number of consumers might be influenced by the availability of privacy protections, real competition over privacy could result.

Conclusion

It is difficult to say whether the law is really an effective device for protecting privacy. Different attempts have produced a mixed bag of results. In the United States, a broad-based privacy law has been imposed on the federal government, with decidedly uneven results. Other more narrowly focused efforts aimed at limiting the federal government have been constantly weakened. For example, the provision of the Tax Reform Act of 1976 that provides for the confidentiality of tax returns has been repeatedly amended to allow more non-consensual disclosures.[54] The Right to Financial Privacy Act, intended to make it more difficult for the federal government to obtain bank records about individuals, was a weak law as passed, and later amendments significantly undermined its weak protections.[55] Privacy restrictions in wiretapping laws, rooted more firmly in constitutional principles, have been more successful, but these laws too are subject to constant pressures for change.

Beyond the scattered federal statutes, other legal responses to privacy concerns abound. There are federal and state constitutional provisions, a few state laws, tort remedies, and some largely unexplored contractual devices. Some of these have proved effective at times. The diverse, uncoordinated nature of privacy law makes any broad evaluation difficult.

The problem here is not a failure of policy. Privacy policies largely developed in the United States have been adopted and implemented around the world. The code of fair information practices identifies the issues that must be addressed in any privacy law. Nor do we necessarily face a failure of law or legal institutions. Laws regulating conduct can work, and some privacy laws have accomplished their objectives even though some of those objectives are narrowly defined.

The main problem is that effective fair-information-practices laws require active, regular oversight and enforcement if they are to be effective. Without some natural oversight—as provided by the courts in wiretapping laws—the laws are likely to have only limited effects on the activities of bureaucracies, governmental or private. Enforcement through individual action does not work well.

A second problem is that the target is so broad. Many different institutions, public and private, directly affect personal privacy interests. This makes the political and regulatory challenge especially difficult.

A third problem is that the target keeps changing. New types of personal records and new record keepers emerge constantly. Even where there is a law protecting a class of records, new technology or social needs may create similar records that fall outside the scope of the existing law. For example, the period 1985–1995 saw the emergence of many new major institutions that use identifiable medical records to carry out statutory or other socially useful activities.

The underlying issue for creating formal and enforceable fair information practices may be one of incentive. Privacy principles have generally not been implemented in ways that offer natural incentives to record keepers to comply. Few existing legal devices have proved effective in pressuring record keepers to take affirmative steps to meet privacy objectives. If adequate pressure or interest exists, any of the devices may work.

Statutes, torts, and contracts could be more effective if there were either an especially attractive carrot or a realistically threatening stick.

In the end, the complexity of the privacy issue and the multiplicity of legal responses and jurisdictions may be the key to a more rational approach. Traditional distinctions between types of records and categories of record keepers are eroding. Personal data from multiple sources are being merged through computers and computer networks. Information comes from government files, consumer transaction records, corporate files, and international sources as well. Records from some but not all of these sources come with privacy rules. These rules may not only be different; they may conflict.

Computers make it easier to manipulate data, but they require more consistent rules for the use of data. Growing confusion among record keepers about what law or policy to follow may create the biggest pressure in the future for more unified and organized responses. American corporate resistance to privacy statutes may ultimately be worn down by uncertainty. International privacy pressure may be especially useful in this regard. As record-based activities expand internationally, record keepers will struggle to comply with broader and more elaborate foreign laws and policies.

Surprisingly, then, the demand for clearer privacy rules may come in the future as much from record keepers as from record subjects. For example, current industry support for federal medical privacy legislation is based largely on the need for uniformity in today's computerized and interstate health-care environment. To meet their need for clear and consistent privacy laws, record keepers may be willing to accept otherwise unwanted privacy restrictions.

The potential proliferation of privacy laws and policies combined with the increasing internationalization of activities involving personal data—as evidenced in part by the growth of computer networks—may ultimately lead in an entirely new direction. A rational and broadly applicable set of privacy principles may be beyond the reach of the law of any single state or nationality. Attempts at common privacy standards, such as the European Union's Data Protection Directive,[56] show significant strains from national desires to preserve local laws. Spiros Simitis, the first data-protection commissioner in the German state of Hesse, sees existing national laws as a serious handicap to common regulation.[57] Simitis also sees the political pressures for accommodating existing laws as a threat to a high level of protection and to the scope of common regulations. In effect, existing national laws may stifle creativity, responses to new technology, and willingness to conform to new international rules.

One response may be the substitution of *private* law for public law. Efforts at international standards offer a useful approach. The Canadian Standards Association has a model code defining the privacy responsibilities of record keepers.[58] This interesting first step toward common rules may still be too general to meet the need for detailed implementation of fair information practices in specific contexts. Another alternative may be more specific privacy codes jointly adopted by merchants and consumers and enforced through private mechanisms. This may be the only practical way to have privacy rules for international transactions conducted on computer networks, where it can be impossible to determine where a merchant or a customer resides or what law is applicable to any transaction or activity.

Legal mechanisms are available to enforce privacy policies. The code of fair information practices offers a comprehensive outline of those policies, but its application in the United States is spotty at best. The problem is less a shortcoming of existing legal devices and more a failure of interest, incentive, and enforcement. If the will for better privacy rules develops, the law can provide a way to accomplish the objectives.

Notes

1. See, e.g., *Olmstead v. United States*, 277 U.S. 438, 572 (1928) (dissenting opinion).

2. See Paul Schwartz, Privacy and participation: Personal information and public sector regulation in the United States, *Iowa Law Review* 80 (1995): 553, 558–563.

3. See, e.g., Priscilla Regan, *Legislating Privacy* (University of North Carolina Press, 1995).

4. 15 U.S.C. §§1681–1688t (1994).

5. 20 U.S.C. §1232g (1994).

6. Public Law 99-508, 100 Stat. 1848-73 (1986).

7. 50 U.S.C. §421 (1994).

8. Alan Westin, Privacy in America: An Historical and Socio-Political Analysis, presented at National Privacy and Public Policy Symposium, Hartford, 1995.

9. Colin Bennett, *Regulating Privacy: Data Protection and Public Policy in Europe and the United States* (Cornell University Press, 1992).

10. 5 U.S.C. §552a (1994).

11. Secretary's Advisory Committee on Automated Personal Data Systems, Records, Computers, and the Rights of Citizens (1973). See generally Robert Gellman, Fragmented, incomplete, and discontinuous: The failure of federal privacy regulatory proposals and institutions, *Software Law Journal* 6 (1993): 199–238.

12. See, e.g., Senate Committee on the Judiciary, Subcommittee on Constitutional Rights, Federal Data Banks, and Constitutional Rights, 93rd Cong., 2d Sess. (1974) (volumes 1–6); Federal Information Systems and Plans—Federal Use and Development of Advanced Information Technology, Hearings before the House Committee on Government Operations, 93rd Cong., 1st and 2d Sess. (1973–74).

13. Public Law 93-579, §2(a)(2)

14. See, e.g., General Accounting Office, Agencies' Implementation of and Compliance with the Privacy Act Can Be Improved (LCD-78-115) (1978) (improvements can be made in how provisions of the Act are being implemented and carried out); General Accounting Office, Privacy Act: Privacy Act System Notices (GGD-88-15BR) (1987) (only 24 of 53 randomly selected system notices were current).

15. 5 U.S.C. §552a(a)(2) (1994).

16. 44 U.S.C. §3507 (1994).

17. 5 U.S.C. §552a(v) (1994).

18. See Department of Justice, Freedom of Information Act Guide & Privacy Act Overview 478-80 (1994).

19. 5 U.S.C. §552a(a)(7) (1994).

20. House Committee on Government Operations, Who Cares About Privacy? Oversight of the Privacy Act of 1974 by the Office of Management and Budget and by the Congress, House Report No. 98-455, 98th Cong., 1st Sess. (1983).

21. For a history of efforts to control computer matching, see House Committee on Government Operations, Who Cares About Privacy? Oversight of the Privacy Act of 1974 by the Office of Management and Budget and by the Congress, House Report No. 98-455, 98th Cong., 1st Sess. (1983); House Committee on Government Operations, Computer Matching and Privacy Protection Act of 1988, House Report No. 100-802, 100th Cong., 2d Sess. (1988) (report to accompany H.R. 4699).

22. Public Law 100-503, 102 Stat. 2507 (1988).

23. General Accounting Office, Computer Matching: Quality of Decisions and Supporting Analyses Little Affected by 1988 Act (1993) (GAO/PEMD-94-2).

24. 5 U.S.C. §552a(e)(10) (1994).

25. Oversight of the Privacy Act of 1974, Hearings before a Subcommittee of the House Committee on Government Operations, 98th Cong., 1st Sess. 226 (1983) (testimony of Ronald Plesser).

26. 5 U.S.C. §552 (1994).

27. See Department of Justice, Freedom of Information Case List (1994).

28. 15 U.S.C. §§1681-1688t (1994).

29. The Consumer Reporting Reform Act of 1993—S.783, hearing before the Senate Committee on Banking, Housing, and Urban Affairs, 103rd Cong., 1st Sess. 25 (1993) (S. Hrg. 103-247).

30. Video Privacy Protection Act of 1988, 18 U.S.C. §2710 (1994) (restricts disclosure of information about specific selections of customers without consent; limits law enforcement access to identifiable customer information; requires customer notice of law enforcement subpoenas).

31. Cable Communications Policy Act of 1984, 47 U.S.C. §551(h) (1994) (subscribers must be notified about personal data collected; subscribers have right to inspect and correct personal data; restricts disclosure of personal data).

32. Family Educational Rights and Privacy Act, 20 U.S.C. §1232g (1994) (students and parents may inspect and correct student data; restricts disclosure of student data).

33. 12 U.S.C. §3401 et seq. (1994).

34. 42 U.S.C. §227 (1994).

35. *Griswold v. Connecticut*, 381 U.S. 479, 484 (1965) (citation omitted).

36. 277 U.S. 438 (1928).

37. Id. at 470.

38. Louis Brandeis and Samuel Warren, The Right to Privacy, *Harvard Law Review* 5 (1890): 193–220.

39. 389 U.S. 347 (1967).

40. Id. at 350–351 (footnotes omitted).

41. 429 U.S. 589 (1977).

42. Id. at 605.

43. Compare *Slayton v. Willingham*, 726 F.2d 631 (10th Cir. 1984) (finding that the Supreme Court has explicitly recognized that the constitutional right to privacy encompasses an individual interest in avoiding disclosure of personal matters) with *J.P. v. DeSanti*, 653 F.2d 1080 (6th Cir. 1981) (failing to interpret *Whalen* as creating a constitutional right to have all government action weighed against the resulting breach of confidentiality) with *Borucki v. Ryan*, 827 F.2d 836 (1st Cir. 1987) (*Whalen* appears to have specifically reserved decision as to whether there is a constitutionally rooted duty of nondisclosure regarding personal information collected by the state under assurances of confidentiality). See also *United States v. Westinghouse Electric Corp.*, 638 F.2d 570 (3rd Cir. 1980).

44. 425 U.S. 435 (1976)

45. 988 F.2d 1344 (4th Cir. 1993).

46. The plaintiff also objected to the state's failure to comply with the requirements of the federal Privacy Act of 1974, 5 U.S.C. §552a note (1988), that certain disclosures be made when states collect social security numbers. The state did not comply with this requirement, but it is not material here.

47. 5 U.S.C. §552a(v) (1994).

48. David Flaherty, *Protecting Privacy in Surveillance Societies* (University of North Carolina Press, 1989), p. 305.

49. William Prosser, Privacy, *California Law Review* 48 (1960): 383–423.

50. See James Maxeiner, Business Information and "Personal data": Some common-law observations about the EU draft Data Protection Directive, *Iowa Law Review* 80 (1995): 619, 622. ("Common-law privacy rights are not intended to be a response to privacy issues raised by commercial information processing activities generally. They hardly could be. They mandate no affirmative obligations, such as obligations of notification, data quality, information subject access, or security.")

51. Prosser, Privacy, p. 394.

52. Schwartz, Privacy and Participation.

53. Privacy Protection Study Commission, Personal Privacy in an Information Society (1977), p. 305.

54. 26 U.S.C. §6103 (1994). This provision was amended at least eight times between 1977 and 1993 to broaden the use of "confidential" tax- return information.

55. 12 U.S.C. §§3401–3422 (1994). This act was amended four times between 1986 and 1992 in ways that weakened the privacy protections in the original law.

56. Directive 95/46/EC of the European Parliament and of the Council on the Protection of Individuals With Regard to the Processing of Personal Data and on the Free Movement of Such Data, 1995 O.J. (L 281/31) (November 11, 1995).

57. Spiros Simitis, From the market to the polis: The EU Directive on the Protection of Personal Data, *Iowa Law Review* 80 (1995): 445–469. ("However, while at first the national laws may appear to be a valuable aid in establishing a common regulation, in reality they constitute a serious handicap. Experience has shown that the primary interest of the Member States is not to achieve new, union-wide principles, but rather to preserve their own, familiar rules. A harmonization of the regulatory regimes is, therefore, perfectly tolerable to a Member State as long as it amounts to a reproduction of the State's specific national approach.")

58. Canadian Standards Association, Model Code for the Protection of Personal Information (CSA-Q830-96), 1996.

8

Generational Development of Data Protection in Europe

Viktor Mayer-Schönberger

In Europe, since the 1970s, "data protection" has become a household word used to describe the right to control one's own data. Data-protection norms are by now established and accepted components of the European nations' legal frameworks. But the connotations associated with "data protection" have shifted repeatedly and substantially, and further defining the term turned out to be a futile if not tautological quest.[1] In addition, the term itself is, as many have noted, ill-chosen. It is not "data" that is in need of protection; it is the individual to whom the data relates.[2]

On the other hand, leaving behind completely the familiar term "data protection" is no solution either. One would lose too much: a revered image, an established metaphor that has proved to work in public discourse. So Europe adheres to it while acknowledging that what it is dealing with is, and will remain, a moving target. But in order to understand the European conception of "data protection" one is better advised to put aside theoretical definitions and to use the term in a more dynamic and evolutionary spirit.

To be sure, data-protection laws have been enacted in the vast majority of European nations since 1970.[3] Not only do they signify the awareness of both politicians and the public to the problem of informational privacy; they also evince the dramatic technological changes in information processing.

Much ink has been spilled over the problem of data protection and how it should be dealt with.[4] Most of the work done in Europe has focused on national data-protection norms and their specific implications. Comparative studies are mainly used to support requests for legislative reform of national data-protection norms. The comparatively small amount of truly international work has come from the Organization for Economic Cooperation and Development (which broke new international ground in 1981 by publishing its Guidelines on the Protection of Privacy and Transborder Flows of Personal Data[5]), the Council of Europe, and, quite recently, the European Union. In Europe, almost all the national norms enacted after 1981 reflected the spirit if not the text of the OECD guidelines.[6] It is difficult, though, to accurately measure the actual, genuine influence of those guidelines. After all, the principles embodied in them were themselves distilled from already-enacted data-protection norms in Europe.

In the early 1980s a European Convention on Data Protection sponsored by the Council of Europe was signed, but it has had little practical impact on national discussions.[7] Belatedly, in 1995, the European Union, after almost a decade of discussion, passed legislation directing its member states to enact specific data-protection norms.[8]

Thus, data protection has been seen largely as a national issue embodying unique national traits of privacy and individual self-determination. Until recently, most of the comparative studies of data-protection norms emphasized cross-border dissimilarities and ventured to explain these variations as evidence for different country-specific perceptions of the data-protection issue. Lately, this national outlook has come under increasing pressure from the mounting international demand for massive cross-border flows of information and the resulting desire for a more homogeneous European data-protection regime.

In addition, a recent preliminary study focusing on various aspects of privacy and data protection and on their interrelation relation failed to confirm predicted differences among nations as diverse as the United States, Thailand, Denmark, the United Kingdom, and France.[9] If significant differences were detected at all, they had more to do with different aspects of privacy and data protection within one country than with similar aspects across countries.[10]

The empirical results of the aforementioned study only reinforce Colin Bennett's[11] eloquent demonstration that data protection, above and beyond national idiosyncrasies, can be viewed as an informally coordinated international process in which nations might be at different stages of legislative development but cannot resist a general evolutionary trend within data-protection norms (especially in Europe).

If it is indeed the case that data-protection statutes develop in an informal but loosely coordinated international context, one should shift away from examining differences among various national data-protection frameworks. It might be more helpful, particularly in Europe, to look at data-protection norms not by analyzing national distinctions but by grouping together similarities of the various data-protection regimes. Legislative development of data-protection statutes among the European countries might become more visible and understandable if seen as a continuous process in which, over time, certain models of data protection surface, prevail, and ultimately wither.

This chapter will attempt to describe the development of European data-protection norms in such generational terms. Through the generational model the recurring themes in the debate over data protection can, it is hoped, be better understood. To be sure, such a generational approach cannot render exact results. Comparing norms of different legal systems always requires some simplification. But absolute exactness is not necessary. Any tension caused by the categorization of norms into distinct groups only fosters holistic analysis.

First-Generation Data-Protection Norms

The first data-protection laws were enacted in response to the emergence of electronic data processing within government and large corporations. They represent attempts to counter dim visions of an unavoidably approaching Brave New World exemplified by plans discussed in the 1960s and the early 1970s to centralize all personal data files in gigantic national data banks.[12]

The data-protection law of the German state of Hesse (1970),[13] the Swedish Data Act (1973),[14] the data-protection statute of the German state of Rheinland-Pfalz (1974),[15] the various proposals for a German Federal Data Protection Act,[16] the Austrian proposals for a Data Protection Act (1974),[17] and the German Federal Data Protection Act 1977[18] can all be seen as direct reactions to the planned and envisioned centralized national data banks. In structure, language, and approach, they represent the first generation of data-protection norms.

To accurately analyze these norms one has to understand the pervasive organizational changes that were taking place during that time. The computer, originally designed to track missiles and break secret codes, came at exactly the right time for government bureaucracy. European nations had just initiated massive social reforms and extended their social-welfare systems. Risks and duties of the individual citizen were delegated more and more to society at large. But this shift in responsibilities required a sophisticated system of government planning, and planning requires data. Thus, government bureaucracies had to constantly collect increasing amounts of information from the citizens to adequately fulfill its tasks and to appropriately plan for the future. But the gathering of data alone is not sufficient. Data must be processed and linked together to create the necessary planning instruments, and so that complex social legislation can be applied to the demands of individual citizens. Hence, in modern social-welfare states data processing is necessary in two directions: from the bottom up (to create aggregated planning information out of millions of personal data items) and from the top down (to transform the perplexing social regulations into concrete individual entitlements).

Without computers, a modern welfare state could not operate. This explains the thinly veiled euphoria of the bureaucracy for the new technology. But not only governments immediately understood the benefits of the computer. Large corporations could better plan, administer, and manage their enterprises, too. This created fertile soil for gigantic proposals to nationally centralize information. In Sweden, census and registration records had already been merged, and tax data was stored in centralized tax data banks. In the late 1960s Sweden's legislature proposed to merge all these information sources into one national information bank.[19]

Similar plans existed in Germany. The state of Bavaria wanted to use the computers installed for the Olympic Games in 1972 for a centralized Bavarian Information System.[20] Hesse's 1970 plan suggested the massive use of centralized information processing in the Hessian state administration.[21] On the federal level, Germany created a special coordination committee to link the planned municipal, state, and federal data banks into one all-encompassing system.[22]

Resistance against such monstrous proposals swelled. The citizens' fear of an automated and largely dehumanized bureaucracy produced a uniting force spanning borders and linking data-protection movements. Technology was at the center of this critique, and the fear of total surveillance by an electronic Big Brother fueled it. This attitude toward centralized data banks echoes in the structure, the language, and the approach of the first-generation data-protection statutes.

Most of the first-generation data-protection norms do not focus on the direct protection of individual privacy. Instead they concentrate on the function of data processing in society. According to this analysis, the use of computers itself endangers humane information processing.[23] Data protection is seen as a tool specifically designed to counter these dangers. The computer is the problem, it seems, and its application must be regulated and controlled.

Consequently, the first-generation data-protection norms take a functional look at the phenomenon of data processing. If the act of processing is the actual problem, then legislation should target the workings of the computer. Data-protection norms were seen as part of a larger attempt to tame technology. If technology is a powerful tool, it must be used as a forceful means of political and social change. The use of data processing, according to this outlook, must be regulated to ensure that it complies with the goals of society at large. Special social and political procedures were devised to ensure the "correct" use of information processing. The inclination of legislatures in the early 1970s to enact functional data-protection norms focusing on processing and emphasizing licensing and registration procedures aimed at controlling *ex ante* the use of the computer stems from this outlook. Hesse's 1970 data-protection statute, the first such law in the world, embodies this functional approach.[24] It deals primarily with the fact of information processing itself (§1), and it regulates the conditions under which data processing may legally take place. In addition, measures of secrecy (§3) and security are prescribed. The Swedish Data Act of 1973 is even more functional by design. Issues of data security, secrecy, and accuracy substantively dominate its functional rules governing data processing (§5 and §6). Adalbert Podlech's 1973 proposal for a German Federal Data Protection Statute contains an entire chapter prescribing organizational rules of processing,[25] security,[26] and source-code integrity.[27] The Austrian government's 1974 proposal for a data-protection law foresaw specific legal regulations to be passed for each data bank established.[28]

Consequently, first-generation data-protection norms do not entrust individual citizens to ensure compliance with substantive data-protection regulation. Instead they set up special institutions to supervise adherence. The data-protection commissioner established by Hesse's 1970 data-protection law,[29] the commission put in place by the 1973 data-protection statute of the German state of Rheinland-Pfalz,[30] Germany's Federal Data Commissioner,[31] and (most prominent) Sweden's Data Inspection Board[32] all are given power to investigate compliance with data-protection norms.

The functional view of first-generation data-protection norms is visible in another aspect of such norms: a number of early data-protection statutes address explicitly the danger of data processing for the balance of power within government. Because the executive arm of a government collects such a multitude of individual data, it holds in its hands a planning and control instrument of enormous power. A legislature, which is supposed to enact laws based on these planning data, lacks such direct information and data access. Consequently, some of the early data-protection norms established functional information and access rights for the legislatures to data collected and stored by the executive branch.[33]

The first-generation statutes avoid using well-known words such as "privacy," "information," and "protection of intimate affairs"; instead they employ rather technical jargon: "data," "data bank," "data record," "data base," "data file." Only "data" in "data banks" is regulated.[34] And "data banks" are required to be registered,[35] sometimes even licensed.[36]

The structure of the data-protection laws was tailored to regulate the envisioned central data centers. Complicated registration and licensing procedures were established. Because a few gigantic data banks were anticipated, linking data-protection norms number to technical context was seen as acceptable and useful. Such a limited number of data banks, it was thought, could be controlled and regulated through special procedures. As these data banks were to be created only after substantial deliberation, data-protection licensing procedures could be started even while a data bank itself was still in its planning phase. In addition, data security at centralized data banks could be maintained by simple physical access controls.[37] Under these controlled and controllable conditions, even the requirement to denominate a specific data-protection official for each and every data bank sounds sensible.[38]

Complex and rich procedures to control and regulate the use of technology took precedence over the protection of individual privacy rights. This created an environment for data-protection laws that were, in approach, language, structure, targeted at a very particular phase in the organizational development of electronic data processing.

But the ambitious plans of centralized data banks were not realized. Not only did citizens oppose them; in addition, technology developed in a different direction during the 1970s. "Minicomputers" surfaced and for the first time allowed small organizational units in government and business to use decentralized electronic data processing. What had originally been a modest number of data-bank providers and potential data-protection violators grew into thousands of individual organizational units of electronic data processors. The image of a monolithic Big Brother who could be fairly easily regulated through stringent technology-based procedures gave way to a broad, blurry picture of a constellation of distinct and novel potential data-protection offenders.

This led to a shift in the data-protection discussion. Data protection, it was argued, should be extended to data processing in even small private businesses and not limited to government administration and centralized data banks,[39] while existing data-protection registration and licensing processes turned out to be too complex and time-consuming in a world with not dozens but thousands of data-processing units. As a result, public administration and businesses started to openly disregard legal data-protection procedures.

In addition, some of the technical terms and concepts used in the first-generation norms (for example, "data bank" and "data file") had lost much of their validity. Consequently, amendments to existing data-protection statutes were suggested.[40]

Finally, many citizens themselves had experienced quite distinctly and directly the potential dangers of unrestricted gathering and processing of personal data. As a corollary to these personal experiences, more and more citizens requested individual privacy and data-protection rights, above and beyond legislative attempts to control and regulate a certain data-processing technology.

The Second Generation: Warding Off More and Different Offenders

Data protection in the second generation focused on individual privacy rights of the citizen. Well-known sources of privacy, such as the right to be let alone, and the right to delimit one's own intimate space, were brought back into the discussion. Data protection was now explicitly linked to the right of privacy, and was seen as the right of the individual to ward off society in personal matters.[41] Informational privacy became a right guaranteed by the constitutions of Austria, Spain, and Portugal.[42] Now the roots of data protection were seen in the negative liberties, in the individual freedom of the people, in concepts stemming from the Enlightenment, the French Revolution, and the American Declaration of Independence.

The peril was not Big Brother. It was no longer manifested in a handful of centralized national data banks that had to be regulated, registered, and licensed from the very beginning. The danger was now seen to lie in dispersed data processing by thousands of computers across the country, and the best remedy was thought to be for the citizens to fight for privacy themselves with the help of strong, even constitutionally protected individual rights.

Data-protection norms created and adapted during this time are at first sight not utterly dissimilar to first-generation statutes. But a closer look reveals modifications. Technical jargon had been eliminated, and definitions had become more abstract and less linked to a particular stage of technology. Existing individual rights were reinforced, linked to constitutional provisions, broadened, and extended. Regulatory procedures emphasized registration over licensing, and some standard data processing was exempted.

The French,[43] the Austrian,[44] to a certain extent the Danish,[45] and the Norwegian data-protection statutes[46] stood at the beginning and in the forefront of this second generation. To be sure, all norms of data protection always included rights of the individual to access and correct his or her personal data. But during the first generation of norms these individual rights were interpreted functionally. They were seen as supporting the accuracy of the personal data stored and processed. Individuals could not decide on whether their data was processed at all; they could merely rectify misleading or inaccurate information about themselves.[47]

In the second generation of data-protection rights, individuals obtained a say in the process. Their consent was sometimes a precondition to the data processing; in other instances, individual consent might overwrite a legal presumption that prohibited processing. These rights differed substantially from the rights to access, modify, and under certain conditions delete one's own personal data. The newly established individual rights delegated explicit decision power to individual to choose what of their personal data would be used for what purposes. For example, §7 and §33 of the Norwegian Data Protection Act delegated to individuals the right to refuse processing of their data for purposes of direct marketing and market research. The Danish data-protection norms entrusted individuals with the rights to decide on the transfer of personal data from public to private data banks, on the storage of sensitive and old information in private data banks, and on the transfer of consumer data.[48]

In addition, the dramatic growth of cross-border flows of information added a new dimension to the national data-protection norms. But the overall approach continued to be national, and international data exchange was handled on the basis of rules of reciprocity.

Thus, rather peculiar data-protection norms were created. Although related to their first-generation predecessors. they nevertheless tried to follow a different path. The hope for grand solutions had been abandoned. Legislatures now hoped—without providing a theoretical or empirical reason for their enthusiasm—that the individual citizen was the best guarantor of successful data-protection enforcement. Data protection as an attempt to regulate technology was transformed into an individual liberty of the citizens.

In Germany, on the federal level, this change of attitude—best and most eloquently expressed by the first Federal Data Commissioner, Hans-Peter Bull[49]—remained largely rhetorical. But the German state legislatures took up the theme in their data-protection laws. The Norwegian, Danish, and Austrian data-protection laws, all enacted in 1978, provided further guidance.

The thematic change is traceable in the institutional domain of data-protection enforcement as well. Now data-protection institutions saw shifts (or, better, additions) to their list of tasks. These shifts were direct responses to the new extra emphasis on enforcement of newly guaranteed and extended individual rights.

First, some second-generation institutions not only investigated data-protection offenses and controlled enforcement but also turned into something like data-protection ombudsmen for individual citizens. When individual rights were strengthened, citizens had to be given an institution to which to report offenses—an institution that somehow helped them in their quest to enforce individual data-protection rights.

Second, some data-protection institutions were transformed or set up to become adjudicatory bodies that rendered opinions on how bureaucracy may or may not interpret data-protection rules. This was the case with the French and the Austrian data-protection commissions. The French commission is not only responsible for the registration procedures of data processing, similar to the Swedish Data Inspection Board; it also functions as an adjudicative body on disputes over individual rights (Article 35). The Austrian Data Protection Commission, a particularly powerful institution, decides as a special quasi-tribunal body on data-protection controversies between individual citizens and the public authorities (§14). The Danish Data Surveillance Authority (DSA) has also power to decide over individual claims of rights violations,[50] similarly to the situation in Norway with its Data Inspectorate.[51]

This re-orientation of data protection from technology regulation to individual liberty and freedom linked it rhetorically with old legal categories of personal privacy. But the noble ideals of negative liberty and individual freedom remained largely political wishful thinking. Their transformation into black-letter law was bound to fail. It is impossible to realize individual informational liberty and privacy without endangering the functioning of the complex European social-welfare states. In real life the individual rarely had the chance to decide between taking part and remaining outside society.

Individual entitlements to social services and government transfer payments require a continuous flow of information from the individual to the government bureaucracy. Citizens and society are so intensely and subliminally intertwined that a deliberate attempt by an individual to resist such information requests, if possible at all, carries with it an extraordinary social cost. Similarly, from bank and money matters to travel and voting, disclosure of personal information more often than not is a precondition to individual participation.

Data protection as individual liberty might protect the freedom of the individual. It might offer the individual the possibility of fending off overboarding societal information requests. But what price does one have to pay for that? Is it acceptable that such data-protection liberties can be exercised only by hermits? Have we reached an optimum of data protection if we guarantee privacy rights that, when exercised, will essentially expel the individual citizen from society?

The Third Generation: The Right to Informational Self-Determination

These and similar ideas have led to a third major reform of data-protection laws. Individual liberty, the right to ward off invasions into personal data, was transformed into a much more participatory right to informational self-determination. The individual now was to be able to determine how he or she would participate in society. The question was not whether one wanted to participate in societal processes, but how.[52]

This was the line of argument of the famous 1983 census decision[53] of the German Constitutional Court that popularized the term "informational self-determination."[54] It is no coincidence that this participatory approach came at a time when civic virtues and traditions, emphasizing active and deliberate participation over negative liberties and freedoms, and were enjoying a sudden revival.[55] This reinterpretation of data protection as a right to informational self-determination can be easily found in the decision of the German Constitutional Court: "The basic right guarantees the ability of the individual, to decide in general for himself the release and use of his own personal data."[56] The court decision not only linked data protection explicitly with a constitutional provision by "finding" a constitutionally guaranteed right to informational self-determination; it also had profound consequences for the entire structure of the originally more functionally oriented German Federal Data Protection Act. The court declared that all phases of information processing, from gathering to transmitting, are subject to the constitutional limitations. Consequently, participation rights of the individual need to be extended to all processing stages. The individual cannot only, as in second-generation data-protection norms, once and for all decide in an "all-or-nothing" choice to have his or her personal data processed, but has to be—at least in principle—continuously involved in the data processing. In addition, the court made clear that the government must, whenever requesting personal data from the citizens, explain why the requested data is needed and what specific consequences the citizen's denial of consent to processing might entail.

The court also found the language of existing participation rights insufficient. Under the functional data-protection regime, information processors only needed to take into account individual interests "worthy of protection." In its decision, the court in effect replaced this objective though indistinct standard with the very subjective personal decision of the individual, to be taken into account when analyzing the legality of personal data processing. The data protection envisioned by the court is archetypal for the third generation. The individual is not under siege in his or her home, pressured to open the floodgates of personal data and deprived of the option to halt them ever thereafter. Rather, the principle of informational self-determination forces data processing to bring the individual human being back into the loop.

The revisions and replacements of the technical first-generation data-protection terms necessitated by the change in technology were part of the second generation. During the third generation of data protection, information technology developed even further away from centralized information-processing models.

Network technology and telecommunication made it possible for workstations and personal computers to be linked together through fast, efficient, and cheap electronic networks. Data is not easily physically traceable anymore, stored in a particular and well-defined central processing unit. Instead, it resides in networks and can be transferred in seconds. Although efficiency is increasing under the network model, the mobility of data contained in it renders futile previously proven technical data security measures.

These changes in context created new legal challenges. But long before, with the second generation of data-protection norms, the legislatures started their retreat from active regulation of technology. Instead of persisting along a difficult path of continuous adaptation of technology-shaping legislation, politicians had chosen to concentrate on more abstract individual liberties and participation rights. The technological tendencies of decentralization only accelerated this legislative flight from regulatory substance.

Thus, data-protection norms of the third generation are characterized by the concentration on—not to call it retreat from—the individual right of informational self-determination and the belief that citizens would exercise this right. They include a hefty portion of pragmatic compromise necessary to bring about legislative results in an often multi-year search for a middle ground between fostering and controlling efficient information processing.

A number of legislative amendments fall into this category: the German states' data-protection statutes in the wake of the Constitutional Court's decision; the late-coming General Amendment to the German Federal Data Protection Act (1990); the 1986 amendment of the Austrian data-protection law, in which licensing and registration procedures were substantially streamlined and deemphasized in relation to somewhat extended individual self-determination rights; the extension of individual participation rights in the Norwegian Data Protection Act; the adoption of a constitutional provision in the Netherlands to guarantee individual data protection; and some parts of the Finnish Persons Register Act of 1987.

The third generation emphasized informational participation and self-determination. The civic republican revival provided only one theoretical underpinning. The other was the more pragmatic legislative understanding that we live in an ever more interlinked and intertwined society in which the individual cannot hope to opt out of participation. Instead individuals must be given a chance to shape the way in which they take part in society through the release of personal information and data. So legislators still believed that the citizens are protecting themselves through guaranteed rights vis-à-vis powerful societal information-gathering and information-processing units. They emphasized the responsibility of the citizens to exercise their rights, but they understood that, to allow the individual to control his or her own informational image, enforcement of individual rights must be stringent and clear.

Individual participation rights were again extended. The various new German data-protection statutes permitted individuals a voice not only in the processing phase but also in every other phase of information gathering, storing, processing, and transmittal. Citizens enjoyed special privileges of participation rights in the field of market research and the deletion of old information.[57] Similarly, the Finnish Persons Register Act contained the right of the citizen to consent to or refuse the transmitting and linking of personal information.[58]

But reality turned out to be different again. Even when empowered with new and extended participatory rights, people were not willing to pay the high monetary and social cost they would have to expend when rigorously exercising their right of informational self-determination. The overwhelming majority feared the financial risk of filing lawsuits and dreaded the circumstances and the nuisance of court appearances. Or (hardly an improvement) they routinely and unknowingly contracted away their right to informational self-determination as part and parcel of a business deal, in which the right itself was not even a "bargaining chip" during negotiations. But, since consent of the data subject had to be sufficient ground to permit information processing if one takes seriously the right to self-determination, such contractual devaluations of data protection were legally valid, and the individual's right to data protection suddenly turned into a toothless paper tiger.

Consequently, data protection, despite deliberate attempts to broaden access and streamline enforcement, remained largely a privilege of minorities, who could economically and socially afford to exercise their rights, while the intended large-scale self-determined shaping of one's own informational image remained political rhetoric.

The Fourth Generation: Holistic and Sectoral Perspectives

The sought-for answers to this development can be amalgamated into a fourth generation of data-protection norms. The legislators realized the generally weak bargaining position of the individual when exercising his or her right. Fourth-generation norms and amendments try to rectify this through two rather distinct approaches.

On the one hand, they try to equalize bargaining positions by strengthening the individual's position vis-à-vis the generally more powerful information-gathering institutions. In essence, such attempts preserve the belief in the ability of the individual to bring about data protection through individual self-determination if the bargaining balance is reestablished.

On the other hand, legislators take away parts of the participatory freedom given to the individual in second- and third-generation data-protection norms and subject it to mandatory legal protection. Such an approach reflects the understanding that some areas of informational privacy must be absolutely protected and cannot be bargained for individually.

Each of these approaches has found a way into the fourth-generation data-protection norms. For example, recent amendments to German states' data-protection laws, and the German Federal Data Protection Statute of 1990, introduced no-fault compensation for individual data-protection claims,[59] thus expanding the Norwegian no-fault compensation model for data-protection claims against credit reporting agencies.[60]

In line with the second approach, certain personal data is taken away from the individual's disposition. Processing of such sensitive personal data is generally prohibited. This concept is manifested in §6 of the Norwegian Data Protection Act, in §6 of the Finnish Persons Register Act, in the Danish data-protection laws,[61] in Article 6 of the Belgian Data Protection Act, in Section 31 of the French Data Protection Act, and in the British Data Protection Act.[62] The Swiss and the German data-protection laws, particularly those of the new German states, do not prohibit processing of certain sensitive data, but instead restrict the contractual bargaining of basic individual data-protection rights to access, correction, and deletion by the individual.[63] Similarly, the 1995 European Union Directive on Data Protection bans the processing of sensitive data (race, religion, political opinions, etc.) except in a few enumerated cases and purposes.[64]

In addition, under fourth-generation developments general data-protection norms are supplemented by specific sectoral data-protection regulations. This signifies the Europe-wide spread and acceptance of an approach originally started in the Nordic countries. The Norwegian, Danish, and Finnish statutes for some time incorporated specific sectoral data-protection regulations.[65] Now even countries in which generalized data-protection norms have a long tradition—including Germany and Austria—are moving toward a general protection framework with sectoral "add-ons."[66] And the 1995 European Union Directive explicitly calls for sectoral "codes of conduct" for data processing.[67]

In the area of enforcement, data-protection statutes now start to establish separate quasi-ombudsman data-protection advocates and more detached, impartial decisional enforcement institutions.[68] While the former take over the role of investigating reported data-protection violations and aiding individual citizens in their data-protection claims, the later assume the adjudicative role of deciding such concrete claims of violation. This is an important and forward-looking step in a number of ways. First, such laws realize the need for both institutions, advocative and adjudicative. Second, this allows for the implementation of various mechanisms to actively support citizens who decide to file data-protection claims. Third, it solves the constitutional problem of mixing together tasks of data-protection advocacy and adjudication. Now the data-protection advocate can actively and publicly push for better implementation of data-protection norms, while the adjudicative body may concentrate on decision making only. The data-protection statutes of Finland (1987)[69] and Switzerland (1992)[70] implement such a bifurcated institutional setting.

With such an institutional framework, the new data-protection advocate is a step away from the concept of ordinary citizens ensuring overall compliance with data-protection rules through their individual claims. This produces an interesting though peculiar symbiosis of enforcement of the individual position and direct state involvement in data processing. The legislators try to rescue the image of an individual's right to informational self-determination. It continues to be at the core of the entire data-protection model. But it is now enforced, detailed, supplemented, and supported. Direct state intervention, the prevalent mode of the functional first-generation models and out of fashion ever since, is partially revived, but relegated to play only a supplementary role in the overall legal framework.

The 1995 European Union Directive on Data Protection, although a compromise document by design, reflects this generational evolution. Individual participation rights rank prominently among this directive's regulations. Accordingly, the directive lists individual consent as one of the legally accepted reasons for the processing of personal data and for the transmittal of personal data to countries with inadequate data-protection regimes.[71] The consent needs to be informed and expressly given in cases of sensitive data processing.[72] Citizens have the right to prohibit the processing of personal data for purposes of direct marketing.[73] Further regulations concern the effective enforcement of individual claims and monetary compensation for established violations.[74] In stark contrast to the data-protection norms of the first generation, which were targeted at taming technology, the directive encompasses not only data processing in computers but also manual data processing in manual files as long as these files are structured and sorted (e.g. alphabetically).[75] Thus, the directive substantially broadens the application of data protection in the administrative domain. In addition, it includes a few specific sectoral data-protection rules.[76]

Conclusion

Since 1970, European data-protection norms have evolved into a dynamic and changing legal framework. While data protection turned into an accepted concept, its content shifted and adjusted to address technological changes and challenges and to take into account philosophical and ideological transformations. Data protection is no longer seen as a purely functional construct to be used to directly shape and influence the use of information-processing technology. Instead, the focus has shifted to the individual. Citizens' rights feature prominently in all European data-protection systems. The individual-rights approach has tended away from simplistic versions of informational privacy as a negative liberty and toward broad participatory rights of informational self-determination, supported and enhanced by a renaissance of direct regulatory involvement. Yet individual rights will remain the centerpiece of data-protection regimes for decades to come.

The evolution of data protection is not over. It is a continuous process. In a couple of years, one might find a fifth or a sixth generation of data-protection norms populating the European legal domain. Though it is difficult to predict the path data protection might take, some general observations can be made.

• The future national data-protection norms in Europe will be much more cohesive than the existing ones. The European Union Directive on Data Protection, aimed at guaranteeing the free flow of data and information throughout the EU while maintaining a high level of individual protection, will drive national legislation toward homogeneity. But technology will do the same. The dramatic growth of the Internet and other supra-national information infrastructures makes the transnational flow of information an everyday matter and thus puts substantial economic and practical pressure on national legislatures to internationally coordinate their regulatory framework.

• Participation rights will continue to be expanded. New information and communication technologies now make it technologically possible to bring the individual into the decision loop when processing his or her personal data. The transaction costs of obtaining individual consent for various stages of data processing from the individual have already decreased and will decrease further.

• The loss of national sovereignty over information flows resulting from internationally interwoven information infrastructures will present a tough challenge to national data-protection enforcement. However, the flexible regimes of participatory rights of the individual, hallmarks of the third and fourth generations of data-protection norms, will allow for contractual structures of dispute resolution to be put in place by the parties involved.

• Technical issues of data protection and data security will continue to be "decentralized" and will face the same fate as the technology they were supposed to regulate. While the rights parts of data protection will remain in general data-protection statutes, technical norms regulating specificity of participation and security will become even more sectoral and particular. The EU Directive on Data Protection points into that direction, and so do the planned sectoral EU directives on data protection for ISDN[77] and for mail order.[78]

• All these legal and technical developments will increase the overall economic and political pressure on the United States to enact similar privacy regulation.

The generational overview has, it is hoped, helped to shed light onto the evolutionary process that has taken place since 1970 and on the two underlying themes ever-present in all the European data-protection rules described: the desire to tame, shape, and instrumentalize technology and the goal of linking data protection with some deeper individual value. Twenty-five years' data-protection legislation has shown national and European legislators what can successfully be regulated and with what consequences. European data protection is not an overcomplex monster devoid of practicality, efficiency, and pragmatism. Instead, it reflects deep convictions and a largely honest attempt to afford effective protection to the only entity in our societies that truly counts: the People.

Acknowledgements

I thank Professor Herbert Hausmaninger for his tutelage and Dean Teree E. Foster for all her help, support, and encouragement.

Notes

1. Sasse, *Sinn und Unsinn des Datenschutzes* (1976), p. 78.

2. The first data commissioner in Germany, Spiros Simitis, noted this fact prominently in the first edition of his commentary on the German Federal Data Protection Law: Simitis, Kommentar zum BDSG (1979), p. 53 and accompanying notes.

3. The German state of Hesse promulgated the world's first data-protection law in 1970. Sweden followed with the first national data-protection statute in 1973. Germany enacted a data-protection statute, the Bundesdatenschutzgesetz (BDSG), in 1977. In 1978, France, Austria, Norway, and Denmark passed their data-protection norms. The United Kingdom followed suit in 1984, Finland in 1987, and Switzerland in 1992. After the fall of the Iron Curtain, data-protection laws were among the first norms to be enacted in the Eastern European countries.

4. See, e.g., Flaherty, *Privacy and Data Protection: An International Bibliography* (1984) and the Information Law Project's data-protection bibliography (printed in Mayer-Schönberger, *Recht der Information* (Böhlau, 1997)).

5. OECD, Guidelines on the Protection of Privacy and Transborder Flows of Personal Data, 1981.

6. The guidelines and their feared economic repercussions had a strong influence in pushing Britain to adopt a data-protection act in 1984.

7. Council of Europe, Convention for the Protection of Individuals with Regard to Automatic Processing of Personal Data, E.T.S. No. 108 (1981), 19 I.L.M. S71 (1981). This convention went into effect in 1985 after France, Germany, Norway, Spain, and Sweden ratified it.

8. Directive 95/46/EG on the Protection of Individuals with Regard to the Processing of Personal Data and on the Free Movement of Such Data, OJ, November 23, 1995, L 281/31.

9. Milberg, Burke, Smith, and Kallman, Values, privacy, personal information and regulatory approaches, *Communications of the ACM* 58 (1995), no. 12: 65–74.

10. The survey looked at four aspects of data protection: the collection problem ("too much is stored"), secondary use, erroneous data, and improper access to data. On a seven-point scale, these five nations varied between 0.3 and 0.8 in their judgment on the importance of any of the four aspects, with the three European nations showing an even higher homogeneity (the maximum deviations were 0.4, 0.6, 0.1, and 0.3, respectively). At the same token, differences

within a specific nation across the four aspects were substantially higher (between 0.7 and 1.3). Ibid., p. 70.

11. Bennett, *Regulating Privacy—Data Protection and Public Policy in Europe and the United States* (1992).

12. Bennett, *Regulating Privacy*, pp. 45–53.

13. Hessisches Datenschutzgesetz vom 7.10.1970, GVBl 1970 I, p. 625.

14. Datalag (Swedish Data Act), dated May 11, 1973. This act was last amended in 1994. For a complete text and translation of the Swedish and most other European data-protection acts, see *Data Protection in the European Community*, ed. Simitis et al. (1992, with updates).

15. Law against the misuse of data of the 24th of January 1974, GVBl. 31.

16. Proposal of a Law to Protect Personal Data Against Misuse during data processing, September 21, 1973, BT-DRS 7/1027; compare Adalbert Podlech's proposal for a much more far-reaching statute in his book *Datenschutz im Bereich der öffentlichen Verwaltung* (1973) and the expert opinion on the proposal by Steinmüller et al., Grundfragen des Datenschutzes, BT-DRS 6/3826, p. 5.

17. Government Proposal of a Federal Statute to Protect Personal Data, Nr. 1423 d. Sten.Prot. NR 13.GP.

18. The full name of the Bundesdatenschutzgesetz can be transliterated as "Statute to Protect Personal Data Against Misuse During Data Processing." The BDSG was enacted on January 27, 1977. See BGBl. I 1978, p. 201.

19. Bennett, *Regulating Privacy*, p. 47; Flaherty, *Privacy and Government Data Banks*, p. 105.

20. Steinmüller, Rechtsfragen der Verwaltungsautomation in Bayern, data report 6/1971, p. 24.

21. Hessische Zentrale für Datenverarbeitung, Entwicklungsprogramm für den Ausbau der Datenverarbeitung in Hessen (Grosser Hessenplan), 1970.

22. Walter et al., *Informationssysteme in Wirtschaft und Verwaltung* (1971).

23. Steinmüller, Grundfragen, in Stadler et al., *Datenschutz* (1975); Simitis, Kommentar, p. 51. See also §1 paragraph 2 BDSG, §1 Austrian government proposal, 1974.

24. Hessisches Datenschutzgesetz vom 7.Oktober 1970, GVBl. I p. 625.

25. Podlech, Datenschutz im Bereich der öffentlichen Verwaltung (1973), §36–§44.

26. Ibid., §45–§50.

27. Ibid., §26–§35.

28. §9 der RV zum DSG, Nr. 1423 d. Beilagen zu den Sten.Prot. des NR 13.GP.

29. Hessisches Datenschutzgesetz §7.

30. §6 Rheinland-Pfälzisches Datenschutzgesetz 1974.

31. §17, BDSG.

32. §15, Swedish Data Act 1973

33. E.g., §6, Hessisches Datenschutzgesetz 1970; §5, Rheinland-Pfälzisches Datenschutzgesetz 1974; Art. III of the proposal for an Austrian constitutional data-protection law, June 11,1974, Nr. II - 3586 der Beilagen zu den Sten.Prot des NR 13.GP.

34. §6 and §2 of the Austrian proposal and §1 of the BDSG.

35. §39 of the BDSG states a duty to register. Similar registration duties can be found in the Austrian proposal of 1974 and in §10 of Podlech's alternative proposal. In that respect the British Data Protection Act of 1984 is somewhat anachronistic, since its registration procedures and its setup are much more in tune with first-generation data-protection norms. The reason for this may lie less in the desire of the British legislators to revive the first-generation data-protection norms than in their inability to envision legislative innovation. Britain passed its data-protection act not because of a desire to tame technology or safeguard the individual, but because business interests feared roadblocks in the flow of data across borders.

36. See §2 of the Swedish Data Act.

37. E.g., §6 and §2 of the Swedish Data Act, paragraphs 1, 2, and 5 of the appendix to §6/1/1 of the BDSG, §9.-(1) of the Danish Public Authorities' Registers Act, §9/2 of the Austrian proposal, and §45 of Podlech's proposal.

38. See the Datenschutzbeauftragter (data-protection official) in the German BDSG of 1977 and the proposals to the Austrian data-protection law.

39. Compare the government proposal for an Austrian data-protection law in 1974, which lacked data protection for processing in the private sector save the weak §24, with the finally enacted version of the Austrian Data Protection Statute in 1978 with an entire section (§17–§31) on data protection for the private sector.

40. Simitis, Kommentar, p. 48.

41. Bull, Datenschutz als Informationsrecht und Gefahrenabwehr, NJW 1979, pp. 1177–1182.

42. Art. 35 of the Portuguese Constitution 1976; §1 (Constitutional Provision) of the Austrian Data Protection Act (DSG) 1978; Art. 18 para. 4 Spanish Constitution 1978.

43. Act Nr. 78-17 on Data Processing, Data Files and Individual Liberties of January 6, 1978.

44. Law of October 18, 1978 on the protection of personal data (DSG), BGBl. 565/78.

45. The Danish statute is still quite strongly influenced by the Swedish first-generation model. Denmark has actually two data-protection statutes: one for the public sector (the Danish Public Authorities Registers Act, Nr. 621 of October 2, 1987) and one for the private sector (the Danish Private Registry Etc. Act, Nr. 293 of June 8, 1978). The translation used for the purposes of this article was published by the Danish Ministry of Justice in October 1987 (no. 622).

46. Norwegian Data Protection Act Nr. 48 of June 9, 1978; translation provided by the Royal Norwegian Ministry for Foreign Affairs, May 1993.

47. As established in §4 of the Hessisches LDSG (1970), in section 10 of the Swedish Data Act (1973), in §11-§14 of the LDSG Rheinland-Pfalz (1974), in §16 of Podlech's proposal, and in §10 and §11 of the Austrian government's proposal of 1974.

48. §16, Danish Public Authorities Registers Act; §4, Danish Private Registers Etc. Act.

49. See e.g. Bull, Datenschutz als Informationsrecht und Gefahrenabwehr, NJW 1979, pp. 1177–1182.

50. §15, Public Authorities Registers Act; §15, Private Registers Etc. Act

51. §8, Norwegian Data Protection Act

52. See Simitis, Kommentar, p. 57; cf. Westin, Privacy and Freedom, p. 7.

53. Decision of the 1.Senate, December 15, 1983, 1 BvR 209/83-NJW 1984, p. 419.

54. Simitis, Reviewing privacy in an information society, *University of Pennsylvania Law Review* 135 (1987), at p. 734; Simitis, Die informationelle Selbstbestimmung, Grundbedingung einer verfassungskonformen Informationsordnung, NJW 1984, p. 398ff; Vogelsang, Grundrecht auf informationelle Selbstbestimmung (1987).

55. For Germany see Kriele, Freiheit und Gleichheit, in *Handbuch des Verfassungsrechts*, ed. Benda et al. (1984); for the US see Karst, Equal citizenship under the Fourteenth Amendment, *Harvard Law Review* 91 (1977), p. 1; Karst, *Belonging to America* (1989). On the civic republican revival in general see Michelman, Law's republic, *Yale Law Journal* 97 (1988), p. 1493; Powell, Reviving Republicanism, *Yale Law Journal* 97 (1988), p. 1703; Sandel, *Liberalism and the Limits of Justice* (1982); Pocock, *The Machiavellian Moment* (1975).

56. Ibid.

57. E.g. §28, (new) German Federal Data Protection Act of 1990; §17, Data Protection Act of German state of Berlin.

58. §18–§20, Henkilörekisterilaki (Finnish Persons Register Act).

59. §7, (new) German Federal Data Protection Act 1990. For an example of a no-fault rule on the state level see §20 of Data Protection Act of the German state of Brandenburg.

60. §40, Norwegian Data Protection Act.

61. But only for the public sector (§9, section 2). In the private sector, individuals must explicitly consent to the storage of sensitive data (§4, section 1)

62. The British Data Protection Act empowers government to further regulate the protection of sensitive data pursuant to section 2.-(3). In addition, processing of sensitive data is restricted by the national data-protection statutes of Ireland, Luxembourg, the Netherlands, Portugal, and Spain. For a helpful com-

parison sorted by issues, see Kuitenbrouwer and Pipe, Compendium of European data protection legislation, in *Data Protection in the European Community*, ed. Simitis et al.

63. See Article 8, section 6, Swiss Data Protection Act (1992); §6, German Federal Data Protection Law (1990); §5, section (1), last paragraph, Data Protection Act of the German State of Brandenburg (B-LDSG).

64. Article 8, EU Directive.

65. §1–§5 of the Finnish Persons Register Act regulate the use of files for research, statistics, market research, direct marketing, and credit reporting. §13 of the Norwegian Data Protection Act regulates the credit reporting, §25 address brokerage, and §31 market research. §8 of the Danish Private Registers Etc. Act restricts credit reporting, §17 direct marketing. This sectoral approach can also be traced back to the very specific approach envisioned by the original Swedish Data Protection Act, which empowered the Swedish Data Inspection Board to issue specific regulations for each file registered.

66. E.g. the sectoral data protection with regard to direct-marketing and list-brokering embodied in the Austrian §268 Abs. 3 GewO 1994, BGBl. 1994/194.

67. Article 27, EU Directive.

68. See Burkert, Institutions of data protection, *Computer Law Journal* 3 (1982): 167–188. Wippermann (Zur Frage der Unabhängigkeit der Datenschutzbeauftragten, DÖV 929, 1994) has shown that the originally advocative German Data Protection Commissioner has become more and more quasi-judicial, and has eloquently traced the start of this shift to the Census Decision of the German Constitutional Court.

69. Finish Persons Register Act, April 30, 1987, 471-HE 49/86.

70. Articles 26 33, Swiss Data Protection Law; §29 and §35, Finnish Persons Register Act.

71. Article 7 (a), Articles 25 and 26.

72. Article 8, section 1.

73. Article 14 (b).

74. Articles 22 and 23.

75. Art. 2. (c) together with Art. 3 (1) of EU Directive.

76. §14 (b), regulating direct marketing; §15 regulating automated decision-making systems; but see the call for "codes of conduct" in Art. 27.

77. ISDN: Integrated Services Digital Network. Directive on the Protection of Personal Data and Privacy in the Context of Digital Telecommunications Networks, to be adopted in 1997.

78. Proposed Directive on Consumer Protection in Respect of Contracts Negotiated at a Distance (Distance Selling).

9

Cryptography, Secrets, and the Structuring of Trust

David J. Phillips

Since the early 1970s, networks, rather than stand-alone mainframes, have increasingly become the standard configuration of computing resources. Data has become vulnerable to unauthorized access, both as it is transferred between computers and as the computers on which it is stored become potentially accessible from any point in the net. In response to the perceived need for greater data security, individuals and institutions have sought to develop cryptographic techniques of data protection and structures and practices that support and integrate these techniques. These pursuits have been extraordinarily successful. Cryptographic techniques have been developed that tax the capacities of even the most powerful surveillance institutions. A particular sort of power is being pursued and contested: the power to keep secrets. These contests involve actors with various interests and with access to various resources. At stake in these contests is the structure of trust: who is trusted to develop and administer systems that regulate the power to keep secrets.

Trust, Power, Secrets, and Cryptography

Trust is a mechanism for proceeding in an uncertain world. At a personal level, in order to go about one's daily business, one trusts that one's perceptions are a dependable guide to action and that, in certain ways, the future will resemble the past. At an interpersonal level, one trusts the guidance and advice of certain friends or family members to lead one through uncertainty. On a social level, trust is "a relationship in which principals . . . invest resources, authority, or responsibility in another [an agent] to act on their behalf for some uncertain future return" (Shapiro 1987, p. 626). Examples of trust relationships include those where principals rely on agents in order to take advantage of the agent's specialized expertise (such as medicine, law, or finance) or to bridge the social distance between mutually distrusting parties (as when an auditor is trusted to publish only certain details of a corporation's financial situation to prospective investors).

By definition, trusted agents are in positions of opportunity to act in ways and in situations where principals cannot. Trusted positions are therefore positions of power, and engaging a trusted agent involves risk to the principal. Principals may act in various ways to avoid or to alleviate this risk. They may opt out of the agency relationship altogether, spread the risk between various agents, or "personalize the agency relationship by embedding it in structures of social relations . . . based on familiarity, interdependence, and continuity" (Shapiro 1987, p. 631). They may also rely on "trustees"—"guardians of trust, a supporting social-control framework of procedural norms, organizational forms, and social-control specialists, which institutionalize distrust" (ibid., p. 635). These social control frameworks may include "ethics codes, standards of practice, regulatory statutes, and judicial decisions [that] prescribe disinterestedness, full and honest disclosure, role competence, . . . or performance consistent with that expected of a 'reasonable person' under the circumstances"(Shapiro 1987, p. 637). They may also include gate-keeping restrictions on agency, such as professional accrediting associations. A critical analysis of this system of trust relationships can serve as a point of entry into the analysis of structures of social power.

The power of trusted agents stems from their access to information or expertise not available to the principals. Trust is necessary only under conditions of uncertainty. Therefore, the management of uncertainty is in part the management of systems of trust. The management of secrets is a special case of the management of uncertainty. Cryptography (literally "secret writing"), or data scrambling, is a technique for managing secrecy. Therefore struggles over the deployment, the oversight, and the management of cryptographic techniques are struggles over the social structure of secrecy, trust, and power.

Table 1
Contested axes of trust: DES and RSA.

	Key length	Availability for scrutiny	Jurisdictional restrictions
DES	Relatively short; fixed	Public, but reasons for development decisions remained classified	Export restricted
RSA	Variable	Public	Only short-key implementations exportable

Table 2
Tactics employed by various actors in contests over DES and public-key systems.

	US federal agencies	Banking institutions	Computing industry	Academic institutions
DES	Development of official standards	Development of infrastructure based on federal standards		
Public-key systems	Use of arms regulations to restrict export and publication		Establishment of patent rights; attempts to promote legislative-branch overview of executive-branch crypto policy	Establishment of prepublication review of crypto research for potential classification

Table 3
Contested axes of trust: Clipper, PGP, and remailers.

	Availability for scrutiny	Access to keys	Jurisdictional restrictions	Identification of users
Clipper	Classified algorithm, tamper-proof chips	Escrowed with federal agencies	US government oversight	Non-anonymous
PGP	Source code available	Managed by users	International	User-created "webs of trust"
Remailers	Source code available	Managed by users	International	Anonymous

Table 4
Tactics employed by various actors in contests over Clipper, PGP, and remailers.

	US government agencies	Crypto patent holders	Private activists	Coalitions of industry and public-interest groups
Clipper	FBI develops and supports laws restricting encrypted telecommunications; administration introduces Clipper as FIPS; Congress pushes administration to open crypto policy decisions; administration abandons Clipper and engages in public-relations campaign		Flaws in Clipper algorithm discovered and publicized; cypherpunks and other activists pursue and receive mass-media attention	Congressional lobbying; Internet-based popular mobilization
PGP	Grand jury investigates PGP-related ITAR violations, returns no indictments	RSA threaten legal action against patent-infringing PGP	Zimmermann creates PGP using patented RSA algorithm, releases PGP source code via Internet	MIT brokers agreement between RSA and Zimmermann; makes PGP available via Internet
Remailers			Cypherpunks develop and maintain remailer systems	

Table 5
Contested axes of trust: SET and Ecash.

	Identification
SET	All details of transaction available to payment processor; keys verified hierarchically
Ecash	Bank cannot identify payer; payer and bank together can identify payee; key verification not specified

Table 6
Tactics used by various actors in contests over SET and Ecash.

	Developers	Banking institutions	US government	Users
SET	Developed by deeply entrenched financial institutions; designed to adhere strictly to current crypto regulation; Congress lobbied to withhold financial regulation	System utilizes extant communications networks and business relationships	Withholds regulatory action	No implementations
Ecash	Developed by non-bankers; attempts to raise public awareness of possibility of private transactions and to reassure government agencies of their continued ability to trace criminal activity; use of patents to control legitimate use	Adopted by small US bank with international market, and by larger European banks	Withholds regulatory action	Develop novel uses for system, sometimes contrary to public-relations intentions of DigiCash

Socio-Technical Negotiations and Social Structuring

Technology is culture made obdurate. It embodies, fixes, and stabilizes social relations. It provides mass and momentum to social systems (Latour 1992). In studying the production and use of technological artifacts, we study the production and use of lasting cultural distinctions and relations. In studying the problem of socio-technical stabilization, we study "the problem of securing the social order" (Law and Bijker 1992, p. 293).

Technological systems not only secure and fix social relations; they are products of social relations. Institutionalized resources, relationships, and courses of action are brought to bear in order to shape technological systems that will in turn constitute and facilitate new institutionalized resources, relations, and courses of action. Technical systems, then, are essentially social, and social systems are essentially technical.

Actors use various resources in order to shape and influence socio-technical change. These may include economic resources, such as patents, market power, or the creation of more cost-effective processes. Resources may also include access to the power of the state, including the power to influence policy making or to instigate police action. Actors may also use cultural resources to influence common understanding or use of systems.

Cryptography[1] and the Structuring of Trust

Three Fundamental Ideas in Cryptography

Trapdoor one-way functions
A one-way function is relatively simple to compute, yet computing its inverse is orders of magnitude more difficult. That is, a function f is one-way if $f(x)$ can be calculated from x in n steps whereas calculating x from $f(x)$ requires n^2 (or n^3, or 2^n) steps. For example, it is easy to compute x^2 in a finite field, yet much more difficult to compute $\sqrt{x}$ (Schneier 1996, p. 29).

A trapdoor one-way function is like a simple one-way function except that the inverse function becomes easy when certain information is provided. For example, the multiplication of two arbitrarily large prime numbers is a one-way function. It is easy to find their product, whereas, given only that product, it is difficult to calculate the two prime factors. If one of the factors is known, though, finding the other becomes a matter of simple division. The multiplication of two large primes, then, is a trapdoor one-way function.

Trapdoor one-way functions are the basic building blocks of cryptographic protocols. Encryption methods generally consist of an algorithm and a key. The algorithm is a one-way function that, when invoked with the original text (or plaintext) and a key used as input, produces the ciphertext. Without the key, the ciphertext cannot be decrypted to reveal the plaintext. With the key, however, the trapdoor of the encryption algorithm can be exploited to invert the algorithm and decrypt the ciphertext.

Thus, there are two levels of security in a cryptographic system: access to the decryption key must be controlled, and the algorithm must be secure. It must be difficult to invert the encryption process without the key. It is not at all easy to prove that encryption algorithms are secure. There is no mathematical proof that one-way functions even exist. The best formal proofs that can be offered are that certain mathematical problems are as difficult to solve as certain other problems that have resisted persistent attempts at solution for hundreds of years. Basing a cryptographic system on this class of problems is one method of making them convincingly secure.

Symmetric cryptography

In a symmetric cryptography system, the same key is used both to encrypt and to decrypt text. In order for Alice to send a secure message to Bob using such a system, they would first agree on a key, then Alice would use that key to encrypt the message and send the resulting ciphertext to Bob. Bob would then use the shared key to decrypt the ciphertext. These systems have two vulnerabilities: First, the shared key must somehow be communicated securely between Alice and Bob. Second, a distinct secret key is needed for each pair of communicants, so as the population of communicants grows linearly the number of keys expands geometrically. Neither of these problems is germane, however, if the encryption system is not used in the communication of messages, but simply for keeping text secure from unauthorized access. In that case, a user's key may be used to encrypt information on that user's own system, and the user need not communicate that key to anyone.

Public-key cryptography

Each user of a public-key cryptography system generates two mathematically linked keys: a private key and a public key. Any message encrypted with the public key can be decrypted only with the private key. To send a message to Bob, Alice would first obtain a copy of his public key. She then would encrypt the message using that key and send the ciphertext to Bob. Bob would then decrypt the message with his private key.

Public-key systems avoid both the problem of the secure communication of keys and the problem of the growth of the number of keys. The private key is never communicated, it remains on the user's presumably secure system. And because each communicant has a single key pair that is used in all communications to him, the number of key pairs grows at the same rate as the number of network users.

Public-key systems also act as digital signatures. Only messages encrypted with Alice's private key will yield under her public key. Therefore, if Alice wants to assure Bob that she is the author of a message, she can encrypt it with her private key and send it to him. When Bob receives it, he attempts to decrypt it with Alice's public key. If the decryption is successful, the message must have come from someone with access to Alice's private key—presumably Alice herself.[2]

The same message may be encrypted repeatedly with various public or private keys, and decrypted in reverse order using the corresponding private or public key, in order to send secure signed messages. For example, to send a signed, secure message to Bob, Alice can encrypt it with her private key, then with Bob's public key. Now it is secure—only Bob can decrypt it with his private key. When Bob decrypts it again with Alice's public key, he is assured it came from her.

Public-key systems are vulnerable to attacks by impostors, however. A malicious party, Mallory, may generate a key pair and issue the public key in Bob's name. Then, when Alice wants to communicate secretly with Bob, she unsuspectingly uses Mallory's public key and sends the message to Bob. Mallory intercepts the message and decrypts it using the private key that he has retained.

Public-key algorithms are relatively slow. In general, they require at least 1000 times as much computing time as symmetric algorithms. For this reason, many cryptographic messaging systems incorporate both public-key and symmetric-key algorithms. First, a unique key (the session key) is generated; then that key is used to encrypt the message using a fast, secure, symmetric algorithm. The slower public-key algorithm is used to encrypt the session key using the recipient's public key. Both the encrypted session key and the encrypted message are sent. The receiver uses his or her private key to decrypt the session key, then the session key to decrypt the message.

Axes of Trust

Trapdoor one-way functions and symmetric-key and public-key implementations are the basic building blocks from which many complex cryptographic systems of secret management are constructed. These systems may be analyzed with reference to several axes of trust. These are the generalized common properties of cryptographic systems whose particular instantiation codifies a certain structure of trust. They determine where secrets are held, who has access to them, who depends on them, and who, in a particular system, is required to trust whom. They are fronts on which social power is contested. Five such axes of trust are length of the keys, availability of the cryptographic algorithm for testing and review, access to the keys, jurisdictional limits, and identification of users.

Key length

An attacker who knows which algorithm was used to encrypt a message, but doesn't have access to the decryption key, may try to decrypt the message by brute force. A brute-force attack is one in which every possible decryption key is tried until one is found that works. In general, the time needed to perform encryption grows linearly with the key length. Adding a bit to the length of the key will add a fixed amount, call it c, to the time needed to perform the encryption. Adding two bits will add $2c$ units of time, adding 100 bits will increase the time by $100c$, and so on. However, adding an extra bit to the key length doubles the number of possible keys, thereby doubling the time needed to perform brute-force decryption. The difficulty of brute-force decryption grows exponentially with key length, while the difficulty of encryption grows linearly.

There is a certain continuum, indexed by key length, in the relative difficulty of encryption and brute-force decryption. For very small keys, encryption and brute-force decryption are both trivial. For somewhat larger keys, encryption is simple but decryption requires significant resources. For still larger keys, encryption is time-consuming and decryption is virtually impossible.

Arguments over key length, then, are arguments over the relative difficulty of encryption and brute-force decryption. There is a range of key length where encryption can be performed by anyone with moderate computing resources but brute-force decryption can be done only by those with extremely powerful computers. In this case, the users of the cryptographic system must yield to the possibility that powerful users will have access to secrets not available to those with moderate computing power. In effect, the shorter the key, the more trust is accorded to those with computing power.

Access to cryptographic algorithms and expertise

Cryptographic systems vary in the degree to which their mechanisms are made available for public scrutiny. At one extreme, the systems may be published as source code, which can be compiled and run on the user's own machine without the use of special hardware. The user may even modify the code before compiling it to customize the system to his or her own requirements. At the other extreme, systems may be implemented on tamper-proof chips using secret algorithms.

Most users lack the expertise to evaluate the performance of a technical system. In order to be assured that the system actually acts as promised, and that no alternate trapdoors permitting unauthorized decryption have been included (intentionally or not), users must trust an agent's evaluation. Managing the availability of that system to scrutiny is a means of managing the availability of trusted agents. If no scrutiny is permitted, users must trust the producer's evaluation of the system. Users and producers may negotiate to permit mutually trusted middlemen to scrutinize and evaluate the system without disclosing its operations. With an open system, users may engage any evaluating agent they wish to engage.

In addition to the availability of the algorithms, the availability of the cryptographic expertise with which to create or evaluate them may be negotiated and contested. Research may be funded or not, and the results may be classified or publicized.

Access to keys

Encrypted secrets will yield to the appropriate key. Secure key-management systems are integral to secure cryptographic systems. Key-management systems may be designed and implemented that permit law-enforcement agents, employers, or system owners to have access to keys. Such a system would not permit users to keep secrets from those agents. On the other hand, systems may be implemented that permit users to control access to their keys and thus negotiate the degree to which they relinquish their trusted position.

Jurisdictional trust

Cryptographic systems are embedded seamlessly in global socio-political systems. These are divided into various jurisdictions, such as nations, states, and treaty organizations. Within and among these jurisdictions, various actors are able to deploy various resources in order to influence the management of cryptographic systems. For example, a law-enforcement agency may be able to subpoena the private key of a user if the user resides in the United States, but not if the user resides in other territories. Therefore, designers and users of cryptographic systems may attempt to limit their deployment and use to jurisdictions in which they have some influence, or whose agency they trust. Simultaneously, other users and designers may attempt to disperse their deployment in order to strategically distribute trust.

Anonymity or identification

The development of the standards of identification incorporated into systems of encrypted communication are negotiations over trust. These negotiations occur in cryptographic systems of several types.

A public-key system requires that messages be encrypted with the recipient's key. The sender, in order to defend against an impostor, needs some way of ensuring that the key he or she is using is in fact the key of the recipient. Yet recipients may wish to remain anonymous, or to maintain a number of personae, each with a different public key, and to control the linkages between these personae. Public-key management systems act as trusted mediators between senders and recipients to certify the link between individuals and their public keys. The degree to which these mediators require identification and the conditions under which they will release identifying information are subjects of negotiation.

Negotiations over anonymity and identification are also reflected in the development of electronic systems for monetary transactions. In one type of system, these transactions may take place by debiting one user's account while crediting another's. In this case, a trusted third party must maintain a link between the individuals and their accounts. On the other hand, transactions may take place through the use of digital bearer certificates—encrypted electronic tokens that act as carriers of value while making no reference to the identity of the bearer.

Identification practices tend to favor organized collectors of information over the identified individuals, since it affords those organizations the opportunity to collate bodies of knowledge about the population while withholding that knowledge from the individuals who constitute that population.

Examples of Socio-Technical Negotiations

The Data Encryption Standard[3] and the Rivest-Shamir-Adleman Algorithm

The first example of socio-technical negotiation involves general-purpose cryptographic systems. Specifically, it examines the creation in the early 1970s of the Data Encryption Standard (DES), which was soon followed by the invention of the Rivest-Shamir-Adleman (RSA) algorithm. The contested axes of trust in this case are the length of the key and the accessibility of the system to expert scrutiny. The DES is a symmetric cipher with a 56-bit key, developed and certified by the US government. The RSA algorithm was the first efficient and reliable implementation of a public-key system. It can be implemented with keys of any length, and it is widely published in source code. The deployment of these systems was contested by various factions in the US federal government, and by the international banking industry, by the US computer industry, and by some academic institutions.

In the late 1960s, the banking industry was increasingly using electronic networks for interbank transfers of funds and in the operation of automatic teller machines. It was also increasingly aware of the vulnerability of these systems to attack.[4] Spurred by the perception of a market need, IBM set up a cryptography research group. Universities, too, established research programs in cryptography. Private firms' interest in cryptography threatened the hegemony that the military had enjoyed over cryptographic expertise.

The US National Security Agency (NSA) was created by a presidential directive in 1952 as a division of the Department of Defense. Although the directive is still secret, the NSA's duties eventually were revealed to include "designing codes to protect US government information, intercepting and deciphering foreign communications, and monitoring international messages to and from the United States" (Pierce 1984). It is a spy agency, and its central mission has included the development and deployment of cryptographic technology. The NSA has vast resources. In 1979 it was the largest employer in the American intelligence community. In 1982 its headquarters included at least 11 underground acres of computing equipment (Bamford 1982). In 1987 its budget, though classified, was reported to be $10 billion (Weiner 1987).

Protecting and intercepting information—secrecy and surveillance—are the NSA's missions. It uses its considerable resources to gather and to withhold knowledge, and to position itself as a nexus of trust such that others have no choice but to rely on its expertise. There is little oversight of its operations.

In 1973 the US National Bureau of Standards (NBS) issued a call for proposals for a civilian encryption standard. IBM had almost finished developing Lucifer, a technique that used a 128-bit key and a series of eight "S-boxes" (digital circuits that performed the actual data permutations). The NBS sought the advice of the NSA on the Lucifer scheme, and the NSA modified Lucifer in two ways: the key was shortened to 56 bits, and the S-boxes were redesigned. The NBS incorporated these changes into Lucifer and published the system as the Data Encryption Standard. In 1976, it was officially adopted as a federal standard.

In 1976 Whitfield Diffie and Martin Hellman published "New directions in cryptography." This marked the invention of the public-key cipher.[5] In 1978 Ronald Rivest, Adi Shamir, and Leonard Adleman published "A method for obtaining digital signatures and public-key cryptosystems." This article included a description of the RSA algorithm, which its inventors later patented.

The contest over the DES and the RSA algorithm focused on key length and on the accessibility of the system to expert scrutiny. Critics of the DES claimed that shortening the key from 128 to 56 bits made the cipher susceptible to brute-force decryption. However, this brute force would be possible only with massive computing resources, such as those available to the NSA. Diffie and Hellman estimated that, with a 56-bit key, a special brute-force decrypting machine, capable of finding a key in less than half a day, could be built for $20 million. Depreciated over five years, this results in an average cost of $5000 per key. Had the key remained 128 bits long, the cost per key for brute-force decryption would be $200 septillion (Bamford 1982, p. 438).

The NSA's attempts to control scrutiny of cryptographic systems have included both controlling access to the systems themselves and controlling the expertise with which to examine them. The agency has pursued this control through classification of information, through management of cryptographic research, through restrictions on the exporting of cryptographic systems, and through influence in the process of federal standard setting. It has been adamant in opposing the dissemination of information about strong cryptography. The NSA's cryptographers would not discuss their reasons for shortening the DES key, nor would they publish the criteria under which the S-boxes were redesigned. They have "a long-standing policy not to comment on the strengths and weaknesses of any cryptologic technique," believing that any discussion may either give enemies clues to US encryption schemes or alert them to weaknesses in their own schemes ("The debate over the US Digital Signature Standard" 1992). Just before the publication of Diffie and Hellman's article, the NSA had tried to halt all National Science Foundation (NSF) funding for encryption research and had tried to become the sole funding authority for such research, thus subjecting all researchers to direct NSA oversight. Instead, however, NSF agreed to refer its research to the NSA for possible classification (Pierce 1984; PCSG 1981).

By classifying cryptographic systems as weapons, the NSA enabled itself to utilize the International Trade in Arms Regulations (ITAR) in managing the development of cryptographic systems. In 1977 the NSA used ITAR as a justification for an attempted prior restraint of all publication dealing with encryption. When the US Justice Department found such prior restraint to be unconstitutional, the NSA pushed for a statutory system of prior review of research relating to cryptography. In response to this threat, the American Council on Education convened the Public Cryptography Study Group. This group, whose members were drawn both from the academic community and from the NSA, recommended a voluntary system of prior review of cryptology manuscripts. In its report, the first public discussion of cryptography policy, the group notes that "implementation of this [voluntary] system will require that the NSA convince authors and publishers of its necessity, wisdom, and reasonableness" (PCSG 1981). Apparently the NSA has been convincing, since prepublication review continues (Landau et al. 1994).

The NSA has also used ITAR to restrict the export of cryptographic equipment, algorithms, and devices. In particular, it has used ITAR to prevent the export of software products embedding the RSA algorithm, unless those products implement the algorithm using a very short (40-bit) key. Export restrictions have hampered the interests of US producers of computing systems, requiring them to avoid using the RSA algorithm at all, to limit their sales to the United States, or to develop separate versions of their product for domestic and foreign markets. The NSA asserts that it is "unaware of any case where a US firm has been prevented from manufacturing and using encryption equipment within this country or for use by the US firm or its subsidiaries in locations outside the United States because of US export restrictions" ("The debate. . ."). However, the Digital Equipment Corporation has abandoned at least one project because of these restrictions. It had developed a system to encrypt data flowing to and from DEC workstations. In order to market the product, DEC would have to point out the insecurity of such communications under the current scheme. However, since it could not offer the solution outside the United States, it could see no marketing strategy that wouldn't leave its international customers believing that their DEC systems were insecure. It therefore dropped the project (Landau et al. 1994).

Official standards are obdurate. The establishment of the DES as a federal standard for the encryption of financial information was the seed for the creation of a complex system with great inertia. Banks accepted the DES largely because by accepting an official standard they met a legal threshold for responsible behavior. If they had attempted to develop and implement their own cryptographic system, they may have been liable for its inadequacies. The official standard transfers that liability to the government, which is protected in other ways from liability. Standards also permit interoperability, which leads to cryptographic systems ever more complicated, more far reaching, and more deeply ensconced in systems of law, economics, and social practice. This obduracy has been formidable, even to the NSA. In 1987, after a scheduled review of the DES, the agency announced that it would no longer support the standard, citing its age and the increasing likelihood of its being broken eventually. However, the agency offered no alternative standard. The financial industry placed considerable pressure on the NSA and the NBS. The NBS reversed itself and reaffirmed the DES, but stated that it would not be recertified in 1993. In 1993 the NBS recertified the standard.

The NSA has used its significant bureaucratic power to influence the social structures of trust instantiated in and through cryptographic systems. Through the use of export restrictions, it has been effective in limiting the international deployment of cryptosystems that it hasn't designed. It is not likely (and indeed it would be a violation of the NSA's charter) that these exportable systems are capable of keeping secrets from the NSA. The NSA has, then, maintained a position where it need not trust others. The most effective mechanisms for creating an enduring structure of trust have been export control and standard setting. The acceptance of the DES as a banking-industry standard, and the development of legal, practical, and technical structures based on that standard, imparted to the DES durability and obduracy, and stabilized the social relations those structures mediated.

Clipper, PGP, and Anonymous Remailers

In 1993 the US government again tried to use standard setting as a lever in creating a durable system of trust. This time, it failed.

The Clipper chip, PGP, and anonymous remailers are secure messaging systems. Clipper is an NSA-designed chip designed for encrypting voice messages. It implements a secret algorithm on tamper-proof chips, and it requires the user's keys to be escrowed with the government. PGP is a public-domain implementation of the RSA algorithm that simplifies the encryption of e-mail and implements "webs of trust" for key certification.

Anonymous remailers relay encrypted messages across jurisdictional boundaries to allow senders and receivers to maintain anonymity.[6]

Clipper is the telephone implementation of the more general Escrowed Encrypted Standard (EES). It uses a tamper-proof chip embedded in special secure telephones. Each chip is programmed with two unique numbers: the chip's serial number and its master key. Because the chip is tamper-proof, no one, not even the owner of the phone, can gain access to these two numbers by interrogating the phone. However, when the chips are produced, two databases are kept. Each records the serial number and half of the master key. Each database is escrowed with a separate agency of the executive branch of the US government.

When a call is made, the two parties agree on a session key.[7] Then each chip produces a Law Enforcement Access Field (LEAF). To produce a LEAF, the session key is encrypted with the chip's unique master key. The encrypted session key and the chip's serial number are then both encrypted with a family key common to all chips and available to law-enforcement agencies. The LEAF is transmitted at the start of each Clipper session, then the rest of the session is encrypted using the session key and a classified algorithm known as Skipjack. In order to decrypt the call, the entire session is intercepted and recorded. The LEAF is decrypted by using the family key to reveal the chip's serial number and the encrypted session key. The serial number is presented to the two escrow agencies, each of which surrenders its half of that chip's master key. The reassembled master key is used to decrypt the session key, which is then used to decrypt the session (Froomkin 1995). (See figure 1.)

PGP is an acronym for "pretty good privacy." It is a free and publicly available system for exchanging secure electronic-mail messages. To send a PGP message, the user creates a random 128-bit session key and encrypts the message using that key and the IDEA algorithm, a patented symmetric-key cipher. The session key is then encrypted with the recipient's public key. The encrypted session key and the encrypted message are then bundled and mailed to the recipient, who uses a private key to decrypt the session key and the session key to decrypt the message. The PGP system automates all this, so that the user need only tell the program which file to encrypt for which recipient. Front-end interfaces for various common mailing systems are available that automate this even further, so the user

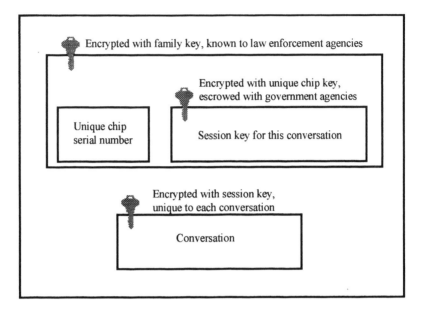

Figure 1
Schematic diagram of Clipper conversation.

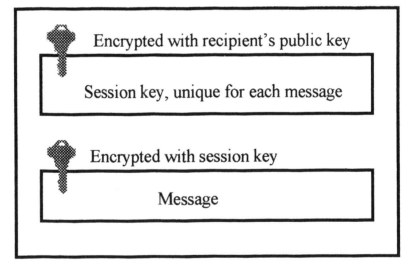

Figure 2
Schematic diagram of PGP message.

need only tell the mailer to apply PGP to the current message and it will automatically be encrypted or decrypted as appropriate. (See figure 2.)

PGP manages public keys through the use of a "key ring." The sender collects the public keys of potential recipients, either by gleaning them from readily accessible databases or by requesting them directly from the potential correspondent. Each key comes incorporated in a key certificate, which includes the key, one or more identifiers for the key's creator (such as name and e-mail address), and the date on which the key was created.

As was mentioned in an earlier section, public-key systems are vulnerable to impostor attacks—e.g., Mallory pretends to be Alice, issues a public key in Alice's name, and intercepts mail intended for Alice. PGP incorporates a method for certifying that Alice's public key was in fact issued by Alice. Each key certificate has an optional field for the digital signatures attesting to the key's veracity. That is, Bob may "sign" Alice's public key by encrypting it with his own private key. If this signed key, when decrypted with Bob's public key, is revealed to be the same key that Alice claims is her public key, then that key has not been altered since Bob signed it. If Carol trusts Bob to sign only the keys of people he trusts, then Carol trusts Alice's key. In this way, "webs of trust" are created where people get their keys signed by many others. Potential correspondents check these signatures to see if any of them are from a trusted party (Garfinkel 1995).

PGP prevents eavesdroppers from discovering the contents of messages, but it still permits eavesdroppers to know who is talking to whom. Remailer systems hide even that information. Suppose Alice wishes to send a message to Bob but doesn't want anyone to know that she is talking to Bob. Two machines, A and B, act as remailers. Each remailer issues a public key, as does Bob. Alice writes a message to Bob and encrypts it with his public key. She then appends Bob's address in plaintext and encrypts the whole with Remailer B's public key. She appends B's address to this, and she encrypts the whole with A's public key. She sends the whole thing to A. Remailer A decrypts the message with its private key, finds B's address and an encrypted message, and sends the message to B. Remailer B decrypts the message, finds Bob's address and an encrypted message, and sends the message to Bob, who decrypts and reads it. Bob knows only that the message came from B; B knows only that it came from A

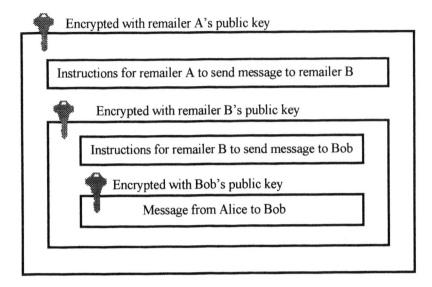

Figure 3
Schematic diagram of message from Alice to Bob via remailers A and B.

and was sent to Bob; A knows only that it came form Alice and was sent to B. Only Alice knows the whole path. This scheme can be altered so that Bob creates a path, encrypts it, and publishes it. Alice sends her message along that path, not knowing its details. Thus, only the receiver knows the details of the path. (See figure 3.)

In either case, a user chooses the path, and the path may include many remailers in many countries and jurisdictions. The system is secure as long as any of its links is secure. This distributed secret sharing ensures that the user need not trust the activities of agents in any particular jurisdiction. The user's secrets are protected unless agents in all the jurisdictions spanned by the system cooperate.

The contestation over these systems aligns about four axes of trust: the accessibility of the systems to expert scrutiny, the accessibility of keys, the dispersal of trust over jurisdictions, and the anonymity of users. Interested actors include the US government, privacy activists and organizations, and holders of cryptographic patents.

As was mentioned above, Clipper is a secret algorithm incorporated in a tamper-proof chip. EES requires that the users' keys be escrowed with two agencies of the executive branch of the US government. Its deployment would establish a structure whereby all users are required to place trust in the cryptographic expertise of the NSA, in the intention of the NSA not to incorporate deliberate trapdoors, and in the protective oversight of the executive-branch bureaucracies of the US government. In short, all secrets are kept in the government, and none from the government. The government trusts no one to act as its agent, and requires everyone to trust. Because it is published in source code, PGP permits its users to trust the cryptographic expertise of any agent they choose. Moreover, PGP permits great latitude in the management of the accessibility of private keys and in the certification of public keys. Users may negotiate their own structures of trust. While Clipper focuses trust on the executive branch of the US government, remailers distribute trust over institutional agents of many jurisdictions. The use of Clipper requires that each communication be identified with the unique terminals involved. The identities of the parties are fairly easily revealed to those with access to the single family key of all the Clipper chips. Remailers permit users to negotiate the degree to which they are identified.

The NSA, the FBI, the executive branch, and Congress have all been active in the negotiations about these systems, as have been industrial interests, traditionally organized public-interest groups, and increasingly numerous and powerful grass-roots activists.

Clipper made its entrance against a background of increased concern among law-enforcement agencies that their surveillance power was being eroded by packet-switched networks that made interception of specific communications difficult and by encryption techniques that made the intercepted communications unintelligible. In 1991 the FBI drafted and promoted a "sense of Congress" clause in an omnibus crime bill. This clause mandated that the manufacturers and providers of electronic communications systems ensure that plain text of the contents of communications be obtainable when authorized by law. Computing-industry representatives, computer scientists, and privacy activists had recently formed coalition organizations, such as the Electronic Frontier Foundation, which had access to significant resources of money, expertise, and energy. These coalitions successfully lobbied Congress to remove the clause from the bill. The clause, though, served as a call to action for many parties concerned with the surveillance powers of the state and with the state's management of encryption systems.

In 1993 the Clinton administration announced EES as a Federal Information Processing Standard (FIPS). FIPS are standards for federal government agencies only; they are formally voluntary for non-government agencies. However, they also become de facto standards for all who interact with the government, especially contractors and providers. Announcing EES as a FIPS, then, was a part of a strategy to use the government's significant market power to encourage production and widespread use of the Clipper system. As a further implementation of this strategy, the Department of Justice immediately bought 9000 Clipper-equipped telephones from AT&T. In a marker of the success of the strategy, AT&T announced that all its new secure telephones would use Clipper. None of the government action was legislative; all of it stemmed from executive-branch decisions (Froomkin 1995, pp. 764–766).

Opposition to Clipper took the form of Congressional persuasion, scholarly refutations of the viability of the algorithm, and mass, grass-roots public action. When the EES FIPS was published for 60 days of comment, only two of the 320 comments received were supportive (Levy 1994). The EFF, Computer Professionals for Social Responsibility, and various industry groups, recently galvanized by electronic-surveillance bills drafted by the FBI, continued their lobbying campaigns, including Internet-generated petitions, to spur Congressional involvement in cryptography standards. In May 1994, Matt Blaze, a researcher at AT&T Bell Labs, discovered and published a scheme for counterfeiting a LEAF field so that it would be recognized by the Clipper Chip but would not, in fact, contain the encrypted session key (Blaze 1995).

The opposition to Clipper was successful. In July, Vice-President Al Gore publicly backed down from EES. In a letter to Representative Maria Cantwell, he stated that, though Clipper was an approved standard for telephone networks, it was not a standard for computer or video networks. Acknowledging industry concerns, Gore welcomed the opportunity to work with industry in developing a key-escrow scheme that "would be implementable in software, firmware, or hardware, or any combination thereof, would not rely on a classified algorithm, would be voluntary, and would be exportable" (Gore 1995, p. 237). Since EES and the Skipjack algorithm (on which Clipper is based) met few of these criteria, Clipper and its associated technologies were no longer supported by government policy. However, the Clinton administration remained committed to key escrow.

PGP, developed in 1990 by Phil Zimmermann, used the patented RSA algorithm. In 1991, Zimmermann asked for and was denied free license to use these patents. However, when Congress began considering the FBI's surveillance clause to the omnibus crime bill, Zimmermann released PGP to various bulletin boards and Usenet newsgroups in a proactive defense against the bill's possible passage. PGP quickly became available globally. RSADSI, the holder of the RSA patent, threatened Zimmermann and further distributors of PGP with legal action. In 1992, RSADSI released RSAREF, a royalty-free RSA library for noncommercial use, to encourage familiarity and the development of commercial, royalty-paying RSA applications. However, because a clause in the RSAREF license made it illegal to use the RSA algorithm if the developer had earlier infringed on RSADSI's patents, Zimmermann still could not legally incorporate the RSA algorithm into PGP. Finally, through the intercession of intermediaries at the Massachusetts Institute of Technology and in the cryptographic community, RSADSI agreed to permit PGP to use RSAREF if a new version were created to be incompatible with the older patent-infringing version. In 1994, such a version was created by Zimmermann and MIT, which "had the strong belief that heavy-duty cryptography, or the ability to encrypt something so that it remains private, needed to be in the hands of the general public" (James Bruce, MIT Vice-President of Information Systems, quoted on p. 106 of Garfinkel 1995).

It is illegal, under ITAR, to export PGP. The MIT server checks that it sends PGP only to machines in the United States. However, all versions of PGP have become available internationally quickly after their release. In early 1993, a grand jury was convened in San Jose to investigate ITAR violations in the export of PGP. For two years, Zimmermann was presumed to be the primary subject of the investigation. In 1995, the US attorney for the Northern District of California announced that no one would be prosecuted in connection with the posting of PGP to Usenet, and that the investigation had been closed. Nevertheless, Zimmermann incurred significant legal expenses, which were offset to some degree by support from EFF and other public defense funds.

Remailers are largely the result of actions by Cypherpunks. "Cypherpunks" is a rubric designating a loose group of computer-savvy people who believe that cryptographic technology has a powerful influence on structures of power, particularly the power of the government to monitor and discipline the economic actions of individuals. In general, they are activists who attempt to implement the technological structures which they believe will alter social structures. Cypherpunks began in 1992 as a loose affiliation of technically adept computer scientists, some of whom had accrued substantial financial assets in the computing industry. Although it began in physical meetings, most communication within the group is through electronic media, particularly the cypherpunks' mailing list. Because it is one of the few generally available sources of cryptographic expertise, the cypherpunks' list has become a source for reporters covering cryptography issues in the general press. As a group, the cypherpunks have available to them access to technical expertise, access to systems and protocols, access to the Internet, and access to the press. They have used these resources to implement working systems of anonymous remailers. However, they have found it difficult to allocate the time and money necessary to keep remailers operational.

In summary, then, actors with various interests, organized in various institutions, used various resources and tactics in the negotiation of the structuring of trust. This negotiation occurred along the axes of key access, jurisdictions, identification, and expertise. No particular tactic or resource was effective in all circumstances.

Federal standard setting, although highly effective in the case of the DES, failed to secure the success of Clipper. The DES was proposed and accepted in the earliest days of non-governmental cryptographic research, when the banking industry had a strong need for a standard and the computing industry had not yet effectively organized to influence public policy regarding cryptography.

Tactics relying on formal legal structures met with mixed success. The federal investigation of Zimmermann may have had a chilling effect on other potential distributors of cryptographic freeware, but no such effect is measurable. Congressional opposition seemed to be effective in persuading the Clinton administration to drop EES, but that too was undoubtedly the result of various and incalculable pressures. ITAR seems effective only against legitimate businesses who wish to be associated with their export products in order to accrue profits.

ITAR is not effective against unorganized, unidentified, anarchic distribution techniques. These distribution techniques are becoming increasingly available, owing to several changes in the socio-technical landscape. Global access to the Internet is making jurisdictional distinctions more and more problematic. It is extremely difficult to limit the international dissemination of information, particularly of cryptographic software. Moreover, the creation of cryptographic software has been recognized as a honorable, political, and potentially profitable pursuit by those with the knowledge and resources to engage in its creation. Hacker culture, the informal collaboration of mutually interested experts, has given rise to a network of network creators, who understand it as their mission to further distributed and secure communications systems. Within this culture, centralized and organized government control is anathema. No government agent is trusted willingly, and socio-technical structures of coerced trust are actively resisted. This resistance has been manifested in the creation and dissemination of software and in the mobilization of public opinion.

In order to influence the cultural context of cryptographic systems, some have attempted to influence public opinion by presenting technology either as exacerbating or as solving social problems. In federal hearings, in the press, and at academic conferences, law-enforcement officers raise the specter of drug dealers, money launderers, and terrorists conspiring via encrypted channels. Louis Freeh, the director of the FBI, has called government control of cryptography a public-safety issue, on the ground that it has the potential to shield child pornographers from investigation (Lewis 1995). Cypherpunks counter that parents should encourage their children to use anonymous remailers in order to protect them from the attentions of pedophiles.

The tactics of many actors also include attempts to make cryptographic systems economically attractive. The Clinton administration relied unsuccessfully on the government's market power to provide an economic incentive for adopting EES. PGP is supported in large part through the resources of MIT. There is as yet no established mechanism for covering the costs of remailer systems. However, cryptographic technique has been applied to the transfer of value over electronic networks. The development of financial cryptography may spur and support the development of all sorts of networked services, including remailer systems.

Financial Cryptography

Two schemes for transferring value over the Internet are in the early stages of development, and it is not clear at all whether either will succeed.[8]

The first of these is the Ecash system, created by David Chaum and the DigiCash company. What is often referred to as "electronic cash" might better be thought of as an implementation of electronic cashier's checks. The cryptographic technique underlying Ecash is the removable signature or "blind signature." To create an Ecash token, Alice first generates a random number and marks it with her removable signature. She then sends the signed token to her bank with the request that the bank debit her account by $10, mark the token as redeemable for $10, and sign the token with the bank's non-removable signature. The bank does so, and returns the token to her. Alice then removes her signature. The token then bears a marker of its value, and the bank's signature, but it can no longer be associated with Alice. Alice transfers the token to Bob in exchange for goods or services. Bob presents the token to the bank, which recognizes its signature and checks that token has not been presented before. If it has not, the bank credits Bob's account with $10.[9]

The second system addressed here is the Secure Electronic Transaction (SET) specification developed jointly by MasterCard and Visa. A draft of this specification was published in February 1996 for public comment. No working implementation of SET yet exists. SET is an attempt to use established credit card payment systems to allow purchasing via the Internet. To begin a SET purchase, the buyer constructs a purchase request. To do this, the buyer elicits from the merchant the public key of the merchant's payment-processing agent ("gateway"). The buyer creates an order information (OI) form and a payment information (PI) form. The PI includes the buyer's credit card number. The buyer encrypts the PI with the gateway's public key, bundles the OI with the encrypted PI, signs the bundle, and sends it back to the merchant, along with the buyer's own

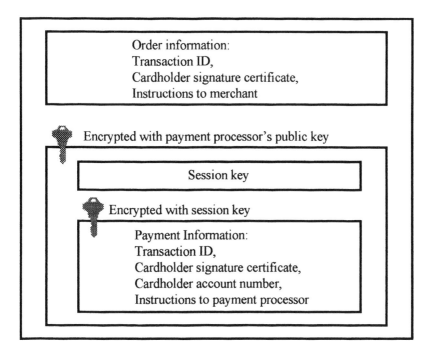

Figure 4
Schematic diagram of SET purchase request from consumer to merchant.

public key. The merchant uses the buyer's public key to check the validity of the purchase request, then forwards the still-encrypted PI to the gateway. The gateway decrypts the PI to discover the buyer's credit card number and, through the established proprietary credit card payment network, authorizes the payment. The authorization is sent to the merchant, who processes the order. The merchant never sees any of the buyer's payment information, though the payment processor may, at the very least, link the purchaser to the time, place, and amount of the purchase. (See figure 4.)

In the SET scheme, key verification is performed through a hierarchical chain. Each public key is sent in the form of a certificate, which is the key signed by a certification authority. The public key of a certification authority is signed by a higher certificate authority. The chain may be of any length. For example, a bank may certify its customers' public keys by signing them with its own public key. The bank's public key may, in turn, be signed and certified by a banking association, whose key may be signed and certified by a government agency or the Post Office. The highest certifying authority makes its public key widely available (it may be published in newspapers of record or distributed over the Internet), so that any tampering is evident. Trust in the highest authority's key is translated into trust in all certified keys through a hierarchical relationship. Before any public key is used, it is verified through the chain of certificate authorities.

These systems differ along the axis of identification. SET relies on a hierarchical certification structure and maintains identification throughout the payment process. Purchasers are linked to their purchases both by the merchant and by the payment processor. SET maintains the identification practices of credit card companies. DigiCash leaves identification policies up to the token-issuing banks, but banks cannot link purchasers with their purchase. The bank knows to whom it issued certain amounts of money, but not where that money was spent. The payer and the bank together can prove who received money, but the recipient and the bank cannot prove from whom money was received. With Ecash, trust is placed in the financial viability of the issuing authority and in the cryptographic strength of the token creating algorithm. In SET, trust is placed in the identification and certification authorities.

Because these schemes build on and impinge on ancient and complex legal, economic, and political structures of international monetary transfer, the parties involved in the negotiations are numerous, varied, and powerful. Among the chief actors in the contest are banks, operators of payment systems, government institutions (including banking and financial oversight agencies, law-enforcement agencies, and tax-collection agencies), merchants, developers of new electronic payment and distribution systems, and political activists.

These systems are embryonic, and the public negotiations about their development are tentative and cautious. Their developers differ in the ways in which they attempt to utilize existing structures of technology, law, and practice in the creation of new structures.

SET builds upon established systems of credit card authorization, payment and billing. At the level of social practice, it depends for its success on consumers' and merchants' familiarity and comfort with credit card purchases, and on a high degree of transparency to the consumer. The purchase ritual is designed to be quick, familiar, and automatic. However, SET's developers have tried to avoid extant structures of legal oversight. In this they have been successful. They have managed to persuade Congress to delay regulation of new systems, and not to define them in such a way that they would fall within the scope of existing regulations (US Congress 1995).

SET is the product of a corporate alliance between MasterCard and Visa, each of which is already strongly aligned and interlocked with banking institutions and each of which has a significant stake in the identification and tracking of individuals through their credit card transactions. DigiCash is a relatively tiny Dutch company, though its marketing plans include the incorporation of its system into US and worldwide banking operations. However, DigiCash perceives its system as a "privacy technology" that breaks the link between identification and transactions and thus empowers individuals with respect to data-gathering organizations. DigiCash is attempting, then, to sell the Ecash system to the very organizations DigiCash feels will be harmed by it. DigiCash is doing this in several ways. First, it is trying to generate and demonstrate a market demand for Ecash by raising public awareness of the possibility of private electronic payments. Second, it is trying to demonstrate that using Ecash is cheaper than handling cash and credit card transactions. Third, it is trying to demonstrate that Ecash's capacity to identify the recipients of payments, and its requirement that each token be immediately redeemed upon receipt, make it an effective tool for law enforcement against money launderers, blackmailers, and terrorists. Fourth, DigiCash is using its patent rights to license Ecash only to organizations that will use it in ways that DigiCash feels will benefit its own interests.

Meanwhile, the success of anonymous remailer networks may depend on their success in incorporating a mechanism for collecting fees for their use. Account-based payment systems, such as SET, are clearly inimical to the anonymity interests of the remailer operators. Ecash might permit a sender to include a payment token in each layer of the remailed message. However, as Ecash is currently deployed, the sender could, in collusion with his or her bank, identify the redeemer of each token, and so reconstruct the remailer chain. Remailer developers are seeking ways to modify

Ecash to permit payee anonymity. However, such a development would probably violate DigiCash's patents and interfere with DigiCash's efforts to place Ecash within standard law-enforcement practices.

Discussion

Users of a cryptographic system are willing to place trust in agents, and to rely on these agents to manage the parts of the system that the users themselves, because of lack of time, energy, desire, or expertise, are not able to manage. They want to trust the agents' knowledge and expertise in the face of their own ignorance and incompetence. They do not want constantly to oversee and negotiate with their agents. Rather, they would like to bestow and cultivate "system trust" (Luhmann 1979, p. 22). They would like to go confidently into uncertainty.

However, this desire for trust is conditional. Users wish, if necessary, to have recourse to certainty regarding the trustworthiness of their agents. They wish to be assured that agents are in fact competent, knowledgeable, and acting in the user's interest. Moreover, they seek not to be fully known to their agents—to retain a degree of privacy in their actions, intentions, and desires. They wish to require that their agents trust them.

Eventually, enduring systems of trust and oversight will be stabilized and incorporated into the non-reflexive actions that seamlessly mediate and reconstitute social practice and social relations. Because stable socio-technical systems fix social relations, actors are likely to invest significant resources to influence the process of stabilization. This chapter has focused on contests and conflicts, both because evidence of these contests will inevitably disappear as the systems stabilize and because similar contests will recur again and again.

Several historical shifts in the availability of social and technical resources have shaped the contest over developing cryptographic systems. Networked communication systems disperse access to expertise and implementations, increase the range of possible degrees of secret sharing, and, in general, destabilize centralized territorial control over systems of secret keeping. As cryptographic expertise becomes widely available outside defense agencies, businesses and financial institutions are less likely to seek assistance from those agencies in designing their systems and thus are less willing to submit to those agencies' attempts to restrict and influence the systems' design. Likewise, widespread expertise means that the developers of systems are less able than they once were to control the degree to which their systems are inspected, adapted, and used. Flaws, bugs, and unintended utilities are discovered and publicized.

Both business organizations and communications systems are becoming global in scope. Businesses resist jurisdictional restraints on cryptographic systems, such as export control or government key escrow, both because the restraints increase the complexity of the businesses' communications systems and because they make it impossible or expensive to sell cryptographic systems globally. However, businesses continue to seek the legitimacy and protection conferred by the state.

Agencies of the US government have sought to retain dominant positions in the structure of trust by restricting cryptographic expertise and implementations to certain jurisdictions, by limiting the lengths of keys, and by requiring central depositories of keys. Whether through "backdoors," through brute force, or through subpoenas, they wish the secrets of others to be knowable. At the same time, they have attempted to keep their own operations secret, both by classifying the details of the development and implementation of cryptographic systems and by promulgating cryptographic policy without public debate. However, the globalization of communication, the expansion of expertise, and the public's interest in cryptographic policy have made these tactics difficult, if not futile. The legislative branch has commissioned its own inquiry into cryptographic policy (Dam and Lin 1996). Though the executive branch has been unable to promulgate effective policy on its own, it remains committed to retaining access to keys (Markoff 1996).

There is more public awareness of the surveillance capacity of businesses and government agencies. Advocates for privacy are at least heard in the press and in Congress. Loosely organized groups of activists have been able to use their technical expertise and the networked communication lines of the Internet to develop and disseminate cryptographic systems that align with the interests of neither business nor government. Their success in mobilizing influence is evident both in the degree to which they have attained a voice in the popular press and in the degree to which business and state interests have opposed them.

Technologies do not succeed on their own. Successful contestation of cryptographic systems, as with all socio-technical systems, will depend on historical circumstance, application of resources, and strategic alignments. Both private business and government interests have the capacity to interfere with and prevent the success of nascent cryptographic systems. The state may deny businesses the ability to conduct their affairs, and businesses may refuse to take up government standards. The Data Encryption Standard is the only government-sponsored cryptographic system that successfully stabilized. It was successful, for the most part, because deeply entrenched financial institutions with significant economic resources took it up and used it in their own interests. If new systems follow this pattern, only those that are supported by both state and economic powers are likely to succeed.

Government and business interests are antagonistic regarding key length, access to expertise, and jurisdictional trust. However, the international scope of expertise, communication, and business organization may force the US government to relinquish its interest in these areas. Government and business interests in access to keys are probably reconcilable, and their interests in identification are aligned. We might expect, therefore, to see systems develop that are provably impervious to brute-force attack. These systems will probably, however, include some bureaucratic form of access to the users' keys, and their users will almost certainly be identified. Strongly anonymous communication may still be available, but it is not likely to be incorporated into deeply institutionalized and commonplace practice.

Notes

1. Much of the description of general cryptographic technique is drawn from Schneier 1996.

2. Actually, the process is less straightforward than this. To save computing resources, a "hash" of the message, rather than the entire message, is signed. A hash function is a standard function that accepts an input string of any length and then creates a much smaller output string in such a way that any change, however small, to the input string results in unpredictable variations in the output string. To sign a message, Alice creates a hash of the message, encrypts the hash with her private key, and sends both the message and the encrypted hash to Bob. Bob is assured that the hash came from Alice, since it yields under her public key. He generates a hash of the message; if it matches the decrypted hash sent

by Alice, he is assured that the message he received is the same message that Alice sent.

3. My description of the DES relies heavily on Schneier 1996.

4. In 1982, the US banking system was transferring $400 billion by computer each day, while estimates of annual losses from computer theft ranged from $100,000,000 to $3,000,000,000 (Pierce 1984).

5. According to former NSA director Bobby Inman, the agency discovered and classified public-key encryption in the early 1970s (Athanasiou 1986).

6. The description of Clipper relies on information from Landau et al. 1994. General information on PGP was drawn from Garfinkel 1995. The description of remailers relies on Cotrell (undated).

7. There are various methods to agree in public on a secret key. That is, both sides of the conversation may be overheard without compromising the secrecy of the agreed upon key. (See Schneier 1996, pp. 513–525.)

8. The descriptions that follow are summaries of "Secure electronic transaction (SET) specification" (1996) and "DigiCash—Numbers that are money" (1996).

9. Ecash is often described using the metaphors of envelopes and carbon paper: Alice puts a blank money order into an envelope. On the cover of the envelope, she writes instructions to her bank to debit her account and to imprint the money order. She signs the envelope. She then puts a piece of carbon paper into the envelope, facing the money order. The bank receives the envelope, debits her account, and imprints the envelope. The carbon paper carries the imprint onto the enclosed money order. The bank returns the envelope to Alice, who discards the envelope and circulates the money order.

References

Athanasiou, Tom. 1986. Encryption: Technology, privacy, and national security. *Technology Review* 89, August-September: 56–66.

Bamford, James. 1982. *The Puzzle Palace*. Penguin.

Blaze, Matt. 1995. Protocol failure in the escrowed encryption standard. In *Building In Big Brother*, ed. L. Hoffman. Springer-Verlag.

Cotrell, Lance. Undated. Mixmaster and Remailer Attacks. URL: http://www.obscura.com/~loki/remailer/remailer-essay.html. Also on file with author.

Dam, Kenneth, and Herbert Lin, eds. 1996. *Cryptography's Role in Securing the Information Society*. National Academy Press.

Diffie, Whitfield, and Martin E. Hellman. 1976. New directions in cryptography. *IEEE Transactions on Information Theory* 22, no. 6: 109–112.

"Digicash—Numbers that are money," February 5, 1996. URL: http://www.digicash.com/publish/digibro.html. Also on file with author.

Froomkin, A. Michael. 1995. The metaphor is the key: Cryptography, the Clipper chip, and the Constitution. *University of Pennsylvania Law Review* 143: 709–897.

Garfinkel, Simson. 1995. *PGP: Pretty Good Privacy*. O'Reilly and Associates.

Gore, Albert. 1995. Vice president's letter to Representative Maria Cantwell. In *Building In Big Brother*, ed. L. Hoffman. Springer-Verlag.

Landau, Susan, Stephen Kent, Clint Brooks, Scott Charney, Dorothy Denning, Whitfield Diffie, Anthony Lauck, Doug Miller, Peter Neumann, and David Sobel. 1994. Codes, keys and conflicts: Issues in US crypto policy. In *Report of a Special Panel of the ACM US Public Policy Committee*. Association for Computing Machinery.

Latour, Bruno. 1992. Where are the missing masses? The sociology of a few mundane artifacts. In *Shaping Technology / Building Society*, ed. W. Bijker and J. Law. MIT Press.

Law, John, and Wiebe E. Bijker. 1990. Postscript: Technology, stability, and social theory. In *Shaping Technology / Building Society*, ed. W. Bijker and J. Law. MIT Press.

Levy, Steven. 1994. Battle of the Clipper chip. *New York Times Magazine*, June 12.

Lewis, Peter H. 1995. The F.B.I. sting operation on child pornography raises questions about encryption. *New York Times*, September 25.

Luhmann, Niklas. 1979. *Trust and Power*. Wiley.

Markoff, John. 1996. White House proposes inititiatives on data scrambling. *New York Times*, July 13.

PCSG. 1981. Report of the Public Cryptography Study Group. *Academe* 67, December: 371–382.

Pierce, Kenneth J. 1984. Public cryptography, arms export controls, and the First Amendment: A need for legislation. *Cornell International Law Journal* 17: 197–236.

Rivest, Ronald L., Adi Shamir, and Leonard M. Adleman. 1978. A method for obtaining digital signatures and public-key cryptosystems. *Communications of the ACM* 35, no. 2: 120–126.

Schneier, Bruce. 1996. *Applied Cryptography: Protocols, Algorithms, and Source Code in C*, second edition. Wiley.

"Secure Electronic Transfer (SET) Specification: Draft for Public Comment," February 23, 1996. URL: http://www.mastercard.com/set.set.htm. Also on file with author.

Shapiro, Susan P. 1987. The social control of impersonal trust. *American Journal of Sociology* 93, no. 3: 623–658.

"The debate over the US Digital Signature Standard." 1992. *Spectrum* 29, August: 32–35.

US Congress. House. Committee on Banking and Financial Services. 1995. The Future of Money—Part 1. Serial 104-27.

Weiner, Tim. 1987. A growing "black budget" pays for secret weapons, covert wars. *Philadelphia Inquirer*, February 8.

10

Interactivity As Though Privacy Mattered

Rohan Samarajiva

Current changes in the economy and in communication, interaction and transaction technologies are seen by some as endangering privacy. The world economic system's transformation from a dominantly mass-production model to a mass-customization model is seen as creating an enormous demand for detailed data on the behavior of consumers. If goods and services are to be customized, it appears necessary for producers and distributors to have access to detailed customer information. Increasing fragmentation of mass audiences also creates demand for data about actual and potential users of specialized media channels.

The current wave of innovation in interactive technologies—the exploding cluster of Internet-related technologies, the upgrading of public wireline and wireless telecommunication networks, and the enhancement of cable systems—is a response to this demand in two ways. First, interactive media systems enable business organizations to better maintain "relationships" with customers. Second, interactive media systems appear to provide an alternative to the increasingly fragmented and decreasingly effective traditional mass media in managing demand. Interactive systems can also collect detailed transaction-generated information[1] (TGI) in a relatively unobtrusive and accurate manner. TGI is increasingly collected and used to predict and modify consumer behavior.

The decline of mass-media industries provides another rationale for TGI. Many see specialized or customized media relying on use-sensitive payment schemes replacing the flat-rate and advertiser-supported payment systems associated with the mass-media model. Even absent demand from firms seeking to build and maintain relationships and to customize products and services, use-based pricing schemes would result in the collection of TGI by media firms. These combined economy-level and industry-level incentives are likely to increase the collection of TGI. If TGI collection occurs without customer consent, privacy, as defined below, will be harmed.

The Surveillance Imperative

The mass-production economy is being gradually superseded by a new economy based on mass customization.[2] While traditional mass production still dominates in volume, the trend is toward mass customization, defined by the use of flexibility and responsiveness to deliver affordable goods and services with enough variety that comes close to satisfying different needs.[3] The trend toward mass customization is especially strong for service-related, communication-intensive industries; however, as Bressand et al. point out, the trend is toward integrating goods and services into complex packages of goods and services or "compacks."[4] Application of information-communication technologies has enabled tighter integration of production and distribution around process technologies that enable relatively low-cost production of a variety of changing products. The old economy was based partly on management of consumption—persuasion of large numbers of people that their needs would be satisfied by identical objects. Since needs are different from person to person, significant persuasion was required to convince people that some mass-produced object would satisfy each person's unique needs. In this economy there was little incentive to find out what each person wanted, because everyone would be sold the mass-produced objects anyway. Audience research techniques of this period worked at relatively large scales of aggregation. Mass production went with mass marketing, and mass marketing went with mass media. There was no reason to offer highly differentiated media content to different groups in society. Concomitantly, there was no reason for media or marketers to engage in fine-grained surveillance of consumers.

The emerging economy allows for flexible production and, thereby, for greater differentiation of products. Profit in the emerging environment comes easier to firms that customize their products and services and position themselves as "high-value" producers.[5] Effective marketing to large numbers of spatially dispersed consumers requires the ability to differentiate prospective consumers and form distinct audiences for marketing messages; to reach these audiences with different persuasive messages about goods, services, or compacks; and to transact with audience members who have been converted to customers. More than in the past, customer loyalty must be retained, because customer "churn" (exit and entry) increases marketing costs. All these actions require the use of information-communication technologies. Going beyond crude segmentation based on areas of residence[6] requires sophisticated surveillance of consumer behavior.[7] In many cases, hitherto discrete anonymous transactions are converted to information-yielding relationships, exemplified by frequent-shopper programs. In sum, mass customization requires the surveillance of spatially dispersed, dynamic target markets and the building of relationships with customers. Customized production goes with customized marketing, which goes with customer surveillance. This is the surveillance imperative.

The emerging economy places great weight on relationships—within the production chain, and between producers and customers. Reliance on one-time transactions is superseded by ongoing relationships. Bressand and others have gone so far as to claim that value is generated in relationships with consumers, who would be better described as "prosumers." What they claim is that consumers' purchase behavior, when fed back to the "manufacturing" firm, directly affects value creation by changing the compack—an example being modulation of toy production based on real-time point-of-sale data by Toys R Us. Consumers are seen as creating value in conjunction with workers on flexible production lines.[8]

Relationships require the parties to know about one another. When small-scale supply of services predominated, proprietors and/or employees of service firms knew individual customers and their preferences regarding a particular service.[9] Indeed, many service-provision activities necessitate relationships with customers.[10] As supply expanded in scale and as the numbers of customers and employees grew large, relationships weakened. However, with increasing use of information-communication technologies, it has again become feasible to "know" customers and their preferences. It is of course debatable whether this actually constitutes knowing, in the sense of one human dialogically engaging with another. Further, this knowing is asymmetrical in two ways. First, the firm knows about the customer, but the customer does not have equivalent information about the firm. Second, the firm in most cases extracts the information from the customer more or less involuntarily in the form of TGI, while the customer has, for the most part, to rely on advertising and other persuasive information disseminated by the firm. The degree of involuntariness ranges from the high end of TGI extraction by monopoly firms holding exclusive franchises (e.g., local-exchange telephone companies)[11] to the low end of competitive firms extracting TGI through point-of-sale routines that require customers to actively resist the extraction of personal information (e.g., collection of telephone numbers for cash transactions by Radio Shack). The value of customer information, particularly TGI, is behind the logic of compacks. Increasingly, firms combine goods and services (e.g., cars with warranties and service contracts) and convert discrete service transactions into relationships (e.g., frequent-travel programs), in both cases, continuing streams of TGI are generated for company databases and for use in consolidating relationships.[12]

Changes in media industries are affected by overall changes in the economy and by internal factors. Decreased demand for advertising to large and undifferentiated audiences affects media industries. There is greater demand for demographic and psychographic data on audiences drawn by various television programs or their equivalents in other media and for ways in which audiences with specific demographic and psychographic characteristics can be produced. With the proliferation of bandwidth to the home in the form of cable and other wire-based systems and direct-broadcast-satellite and other wireless systems, increasingly competitive media channels are unable to deliver mass audiences to advertisers.[13] Because of the clutter of advertising and the changes in viewing habits made possible by remote controls and videocassette recorders, audiences are increasingly expensive to assemble.[14] The inchoate crisis of mass-media industries contributes to the emergence of interactive media systems. As with past changes in media, emergent systems will not be completely independent of "old" media industries, though some new actors may gain footholds in the transition.

The inchoate mass-media crisis creates the possibility of a new pricing paradigm. In the late 1980s, scholarly and industry consensus seemed to be coalescing around use-based pricing. Mosco coined the term "pay-per society."[15] Telephone companies were pushing the public and regulators away from flat-rate to measured-service configurations. Cable companies saw their future in the proliferation of "pay-per-view" channels. The online-database industry had various use-based payment algorithms in place. Flat-rate (e.g., local telephone) and third-party (e.g., advertising-supported mass media) payment mechanisms appeared destined for obsolescence.

The down side of use-based payment schemes was detailed use monitoring. Such surveillance, particularly of information behaviors likely to yield inferences about individual proclivities, has long been considered invasive.[16] The possibility that interactive media systems could aggregate TGI pertaining to financial activities, energy use, and home security compounded this concern.[17] The incentives to surveil for billing would combine with those to build profiles for marketing and other purposes. Increased capabilities of collecting, storing, and manipulating immense qualities of information at low cost would lower the "threshold of describable individuality."[18]

Reports of recent trials of interactive media systems in the United States suggest that these concerns are still valid.[19] Indeed, new proposals for "data metering" portend surveillance of an even more fine-grained form:

> The concept is simple: instead of charging a flat fee for a software program, or an hourly fee for access to a database, data metering allows companies to charge per *use*. . . . Think of this metering device as an electric meter that keeps track of the flow of data into your computer and bills you accordingly. . . . With either system [referring to two commercial systems in development], a user can transfer money onto the meter by providing a credit card number. . . . When a user requests a program off a CD-ROM or an online database, the meter subtracts the appropriate amount . . . and downloads and decrypts the data. Downloaded programs may be set so that they live for only a few days or uses.[20]

Analysis of the Information Infrastructure Task Force's proposals on copyright suggests that the next wave of copyright reforms will incorporate built-in and fine-grained surveillance.[21]

Attenuating Factors

Trust and Privacy

Arrow's assertion that "virtually every commercial transaction has within itself an element of trust, certainly any transaction conducted over a period of time"[22] suggests that trust is an essential ingredient of an iterated set of transactions between the same parties, or a commercial relationship. Giddens, in a discussion of trust in interpersonal contexts, states that "relationships are ties based upon trust, where trust is not pre-given but worked upon, and where the work involved means *a mutual process of self-disclosure*."[23] Both appear to imply that, absent a modicum of trust, a relationship could not exist. It is true that trust-related attitudes are crucial elements of relationships and that trust is the foundation of stable, productive relationships. However, it is misleading to infer that *all* relationships are based upon trust. The corollary to Giddens's claim that trust has to be worked upon is that the work may fail; that instead of trust, mistrust or angst may result.[24] Arrow's claim may hold in an idealized competitive environment; however, customer relationships marked by mistrust and angst cannot be ruled out in the presence of information asymmetries, spatial monopolies, and barriers to exit faced by parties, particularly consumers.

Giddens defines trust as "confidence in the reliability of a person or system, regarding a given set of outcomes or events, where that confidence expresses faith in the probity or love of another, or in the correctness of abstract principles (technical knowledge)."[25] Trust derives from faith in the reliability of a person or a system in the face of contingent outcomes. The primary condition that creates a need for trust, according to Giddens, is lack of full information, generally associated with a person who is separated in time and space or a system whose workings are not fully known and understood.[26] In the presence of perfect information, faith and trust would be superfluous. Lacking complete information about a system, a user has to develop a trust-related attitude toward the system. This can range from complete trust through mistrust to angst.[27] Individuals who interact and transact with complex business organizations lack full information about their workings and have to rely on trust. When such interactions and transactions are mediated by interactive systems, trust is even more important. In view of the documented importance of proximity, co-presence, and talk, particularly for contextualizing interactions and building relationships, interactive systems that minimize proximity heighten the need for trust.[28] Individuals lack information about the business organization and about the interactive system. Businesses also lack information about their current or prospective customers. However, businesses have an alternative to trust. They can obtain more information through coercive surveillance, overt or covert.

Trust is dynamically generated. It has to be "worked on." Trust, both in persons and systems, has strong aspects of mutuality. In interpersonal relationships, one party's actions, particularly self-disclosure or lack thereof, can reinforce, diminish, or destroy the other party's trust.[29] With care, this claim can be extended to customer relationships. Although the symmetry suggested by "opening out of the individual to the other" and "mutual process of self-disclosure" would be unrealistic in customer relationships, some aspects of trust-building behavior (such as absence of coercion in making and receiving disclosures, and a degree of disclosure, albeit non-symmetrical) may be expected from a commercial organization.

These aspects overlap with a definition of privacy derived from research on everyday social practice. Based on studies of how humans interact in public spaces, Goffman stated that "in Western society, as probably in all others, there is the 'right and duty of partial display.' Two or more individuals present together have the right and duty to make some information generally available concerning their relationship and the right and duty to leave unsignaled other information about their relationship."[30] Drawing on this body of work and subsequent scholarship,[31] I have defined privacy as "the capability to explicitly or implicitly negotiate boundary conditions of social relations."[32] This definition includes control of outflow of information that may be of strategic or aesthetic value to the person and control of inflow of information, including initiation of contact. It includes non-release to a third party of information yielded by one party unless with explicit consent. It does not posit privacy as a state of solitude, as suggested by definitions such as "the right to be let alone."[33]

Privacy, as defined above, is situational and relation-specific. In some contexts, a person will voluntarily yield highly personal information and will not consider that release, by itself, a diminution of privacy. In other contexts, the most mundane information will be guarded with great care. The same applies to the reception of information. Privacy is a precondition for trust—an attitude developed on the basis of situational or experiential factors. Trust affects privacy. A user's trust in the information practices of a system is likely to make possible consensual surveillance, which can enhance trust. The resultant spiral will lead to stable and productive customer relationships.

Conversely, where a system engages in coercive surveillance, mistrust and angst are likely to result. Customer mistrust of or angst about the information practices of a system is likely to lead to reduced release of information, tolerance of misinformation, release of disinformation, and greater resistance to receipt of information from the system. In the face of the resultant information problems, the system can either increase reliance on trust or increase reliance on coercive surveillance. The former action is likely to be counterproductive in the short term unless the underlying mistrust or angst on the part of the customer is addressed and removed. The latter action, if known to the consumer, will further decrease trust, giving rise to a spiral of mistrust.[34] If the consumer cannot exit the system, the result will be a customer relationship that is pathological, in the sense that mistrust and angst are dysfunctional to effective relationships. Alienated customers and former customers can communicate their experiences to current and potential customers, thus decreasing their trust—a decidedly undesirable outcome for a business organization.

Outcomes can range from stable, productive relationships to pathological relationships. Actual outcomes are likely to have greater or lesser affinities to the two ideal types. The mere fact of mediation by interactive media systems does not change the outcome. Interactive media systems can be used for consensual as well as for coercive surveillance. The conditions under which a commercial organization will choose to create a trust-conducive environment or take the path of coercive surveillance require independent explanation.

Commercial organizations' knowledge (or lack thereof) of the benefits of stable, productive customer relationships and the costs of pathological relationships may provide a partial explanation. Contemporary commercial practices, which are heavily biased toward coercive surveillance, may be the outcome of ignorance about the results of coercive surveillance and trust-conducive treatment of customers. If this is true, corporate behavior may be expected to change over time in the direction of fostering trust and privacy. However, if commercial organizations mask coercive surveillance through interactive media systems, the current bias toward coercive surveillance may persist. If customers are not aware of surveillance, the destructive spiral leading to pathological customer relationships may be arrested. Pervasive patterns of coercive surveillance may become entrenched and "shift the goalposts" regarding current expectations of privacy and trustworthy behaviors.

In addition, little is known about how to create a trust-conducive environment based on interactive media systems. Trust in abstract systems depends on the "access points" or interfaces of the systems. Assuming that most contacts with abstract systems occur in the form of interactions with experts or their representatives at the access points, Giddens discusses the demeanor of system representatives such as doctors or airline cabin crews as crucial to system trust.[35] By their very nature, interactive media systems minimize routine co-present interactions with humans representing the system.[36] The public's non-volatile and growing concern about privacy, particularly in relation to corporations and interactive media systems,[37] appears to suggest that business organizations are failing to gain trust and indeed are engaging in practices that lead to mistrust and angst among consumers. Indicators of mistrust and angst are becoming evident despite the fact that consumers' knowledge of corporate information practices is incomplete.[38]

The above analysis presupposed a direct relationship between a commercial organization and a customer, mediated by an interactive media system controlled by the commercial entity. The prognosis for trust and privacy may be better where the relationship between the commercial organization and the customer is mediated by an interactive media system controlled by a third party.

Attention

Attention has always been scarce.[39] However, as rapid technological and economic changes accelerate the expansion of electronically mediated forms of communication, interactive and otherwise, the scarcity and value of attention are highlighted. The economist Herbert Simon has stated that "a wealth of information creates poverty of attention and a need to allocate that attention efficiently among the overabundance of information resources that might consume it."[40] Attention is a precondition for the constitution of interactions or transactions (here defined as interactions that result in exchange of value), and, thereby, of relationships (iterated interactions or transactions between the same parties). Attention, once obtained, can be utilized for different forms of persuasion.[41] If the persuasion is effective, transactions and relationships can occur. Not all audiences are markets, but all markets[42] require the production of audiences. Without the assembly of attention, it is not possible for the buyers or sellers to transact. In the marketing literature, the term "prospect" is used to describe the "not-yet-but-likely-to-be" customer. Prospects are better described as an audience, because the latter term privileges the assembly of attention and underlines the commonalties among collectivities attending to political, economic, and cultural messages.

In a complex economy, attention must be produced and reproduced on an industrial scale for use by political and economic organizations.[43] Production of the attention of multiple individuals, not necessarily gathered in one proximate space or at one time, is described as an audience in relation to books, radio, and television.[44] Much of the scholarly work on audiences has been done within the field of mass communication.[45] An audience may be conceptualized in two primary ways. First, it may be seen as an object that is measured, bartered, and sold. Political economists of all persuasions tend to favor this conception.[46] Second, it can be seen as a collectivity of agents who create meaning from media content. Cultural-studies scholars tend to favor this conception.[47] This chapter uses the politico-economic conceptualization while acknowledging the volition of audience members and the relative openness of the meaning-making process. Information-storage-and-retrieval technologies make possible modern, industrial form of audience production and reproduction. Different technological forms yield different structures within which agents make meaning. A book allows for solitary retrieval of information, repeated perusal of text, authorial control over sequencing, and so on. Television, in the classic form, encourages collective viewing, does not allow for review or "freezing" of a part of the content, and so on.[48]

Production, distribution, and retrieval of messages through a medium involve a complex set of institutions (deeply established practices). Different institutions are implicated in different media. Definitions of audiences have implied a commonalty of content that is attended to, as illustrated by this 1855 usage: "'Pilgrim's Progress' . . . has gained an audience as large as Christendom."[49] The explosive growth of technologically retrievable messages (commonly described as a proliferation of channels or as the end of bandwidth limitations) has shaken the traditional conception of audience (and the associated concept of mass communication).[50]

In view of the proliferation of technologically retrievable messages and channels and the emergence of interactive media, it is useful to differentiate among three meanings hitherto included in the term "audience":

An *audience* is defined as the persons attending to a specific message. This stays close to the core meaning of an audience as an auditory collective. Previous scholarly and lay use has been broader. The audience does not have to be in the same place or be paying attention at the same time. Those attending to a message need not reach a common understanding, nor does attention have to be efficacious from the communicator's perspective.

A *meso-audience* is defined as persons likely to attend to a class of messages. A "daypart," a term of art from the television industry, is an example.[51] There is no presumption that all those within a meso-audience will end up in the audiences intended by those who produced the meso-audience. The more effective the process of producing a meso-audience is, the greater the probability of that outcome. However, meaning making is inherently indeterminate.

A *meta-audience* is defined as that from which meso-audiences and audiences may be produced. Uses such as "the audience for Channel 6" or the "the television audience" would fall within this definition. Subscribers to the Internet would be another example. This category is the one that is most relevant to interactive media systems, including telecommunication networks.

This threefold distinction allows the different institutions and incentives associated with each to be identified and analyzed. Meta-audiences, meso-audiences, and audiences are produced and reproduced with effort.

They are fluid and ephemeral. Network operators such as telephone companies are primarily engaged in the assembly of meta-audiences. When AT&T prepares and markets directories of its subscribers organized according to their use of 800 numbers,[52] it is engaged in the assembly of meso-audiences. When a direct-marketing company utilizes such directories in its marketing campaigns, it is engaged in the assembly of audiences. Incentives in terms of attention gaining, surveillance, and relationship building and maintenance differ at each level. A meta-audience is the most basic. Meso-audiences are produced using the meta-audience as a resource. Audiences are produced using meta-audiences and meso-audiences as resources.

Interactive media systems

Producing and reproducing television audiences has never been easy. In addition to such attention-consuming activities as work, interpersonal relations, and leisure, early television-audience producers had to compete with only a handful of other networks or stations in the pre-cable United States, and with even fewer rivals in other countries. The difficulty increased with the entry of cable and direct-broadcast satellite channels, and by an order of magnitude with the wide range of information and activities available on interactive systems such as the Internet. On one hand, producers of audiences struggle to assemble and hold audiences. On the other, audience members struggle to cope with the plethora of objects competing for their attention.

A limited set of devices are used in assembling an audience for a television program: the human attraction to stories, continuity of narratives doled out in installments, "hooks" at the end and at critical points of the installment, stars, scheduling to optimize flow from one audience to another, regular scheduling (daily or weekly), teasers and advertisements, creation of excitement in other media around issues such as "who shot JR?," and so on. Only some of these devices pertain to a discrete message and its audience. Others have to do with the channel or network and the daypart. Meta-audiences are produced by devices intended to draw people to the channel or network; meso-audiences are produced by devices intended to draw people to specific dayparts; audiences proper are produced by devices intended to draw people to specific programs.

To be precise, programs are devices for assembling second-level meso-audiences. The true audience is that which is created for a specific advertisement shown within a program slot.[53] However, depending on the level of analysis, programs with embedded advertisements (overt or covert, as with product placements) can be taken to be the focus of audience attention.

Many of these tried and tested devices are ineffective with interactive media systems, here defined as media systems allowing potential real-time interactivity within the same medium. This definition excludes one-way media (such as radio) that allow interaction via telephone and print-on-paper-via-mail media (such as magazines) that allow for interaction via telephone, e-mail, etc. Actual real-time interaction is not necessary, nor is symmetry of information flows. Ability to complete transactions, including payment, may be included but is not essential. Though the conceptual framework can usefully be applied to hybrid media (such as television shopping channels made up of television, telephone, and credit-card technologies; newspapers that utilize print, voicemail, and/or telephony; and talk radio, which uses radio and telephony), only pure interactive media are addressed here.

Two ideal types may be identified within interactive systems. The first is designed to sell information and/or communication capabilities to subscribers. Here, the revenue stream is dominated by subscriber payments. Standard online services such as CompuServe and the public telecommunication network are examples. An interactive system designed to sell information to subscribers can adopt one of two main pricing schemes: it can sell on the basis of use or it can utilize some form of flat-rate pricing. The former requires fine-grained surveillance. The latter requires less tracking of use, but subscribers may believe that surveillance takes place. Both modes take information, not attention, to be the scarce commodity. In the second ideal type, a system operator sells access to a meta-audience, "universal" or otherwise. Universal is defined as inclusive of all or most households or persons in a spatially defined market.[54] Vendors purchase access to the meta-audience as an input to the process of producing meso-audiences and audiences. Here, payments from vendors (including advertisers) dominate the revenue stream. The US model of commercial television broadcasting is a non-interactive example. The design of UBI, the Québec interactive system currently under construction, is an interactive exemplar.[55]

An interactive system of the second type must attract and retain subscribers and induce them to use the services provided by vendors. In other words, the system operator must enable vendors to produce and reproduce meso-audiences and audiences from the meta-audience. Depending on the concrete conditions affecting consumers, the ways in which these objectives may be achieved differ. Where consumers have few alternatives in terms of allocating attention, the meta-audience producer's task is relatively easy. Where this is not the case, considerable effort, including measures to ensure privacy and build trust, is required.

Interactive systems of the second type do not necessarily have to produce universal meta-audiences. They can choose to produce niche meta-audiences. Here, the dynamics of audience production are less favorable to trust and privacy than with universal meta-audiences. Moreover, niche meta-audiences and meso-audiences tend to be indistinguishable, in that the system operator must provide some reason for a subset of consumers to allocate their attention to one niche system instead of another. In addition, interactive systems (and associated services, such as payment systems) are characterized by first-comer advantages and network externalities. As a result, entrepreneurs will tend to strive for universal-coverage meta-audiences.

Recent work in economics[56] has systematized earlier explanations about why competition does not work very well in certain situations such as competition between videocassette recorder (VCR) technologies or typewriter keyboards. As one VCR format attracts users, more videocassettes are made in that format and more of those videocassettes are stocked by rental and sales outlets. As more people learn to type on a particular keyboard, more typewriter manufacturers adopt that format and more employees require that format. Small increases in the market share of one format improve its market position through positive feedback. Over time, the format with the smaller market share gets driven out, even if it is technically superior. An interactive system with a larger number of subscribing households is likely to attract more vendors. A system with more vendors is likely to attract more subscribers. An electronic wallet system that has the larger number of users is likely to be attractive to more retailers. A system that is accepted at more retail outlets is likely to be more attractive to consumers. In this light, operators of interactive systems and of associated payment systems have incentives to move to be the first to gain a large market share, because that would form the basis of a self-reinforcing set of entry barriers.

A system operator seeking to produce a universal meta-audience is compelled to create a trust-conducive environment. Otherwise, it would not be possible to attract almost all the members of a potential meta-audience to the system, or to enable the production and reproduction of audiences and meso-audiences therefrom. A trust-conducive environment, by itself, will not yield a universal meta-audience. However, since even small groups of privacy-sensitive and trust-sensitive members of the potential meta-audience can prevent universality from being achieved, such an environment is a necessary condition.

A trust-conducive environment requires trustworthy and privacy-friendly behaviors from the system operator as well as the vendors using the system. As a result, the system operator is likely to cajole and even coerce vendors to cooperate in building and maintaining a trust-conducive environment. In view of the incentives all commercial organizations have to build and maintain stable, productive relationships with customers and the difficulties of engaging in covert coercive surveillance on an interactive system designed and controlled by a third party, vendors are likely to cooperate.

Exemplar

UBI is a consortium comprising Groupe Vidéotron (Québec's dominant cable company; 20 percent); Hydro-Québec (owned by the Province of Québec; 20 percent); the Canada Post Corporation (owned by the Government of Canada; 18 percent); Loto-Québec (owned by the province of Québec; 12 percent); the National Bank of Canada (private; 10 percent); Vidéoway Multimédia (a part of Groupe Vidéotron; 10 percent); and Hearst Interactive Canada (a subsidiary of a US media company; 10 percent).[57] The UBI (Universal, Bidirectional, Interactivity) network will draw revenues from vendors or information-service Providers (over 190 have signed up so far) in exchange for access to at least 80 percent of the households in a spatially defined market. Access to the service is free of charge to the first 80 percent of households,[58] who, if they subscribe, will continue to pay for regular and specialty cable services provided over shared network facilities. The cable operator will pay the consortium a monthly fee of C$2.50 per terminal.[59]

Bypassing the technical and economic difficulties of delivering video on demand on a large scale,[60] the UBI system seeks to create an "electronic mall."[61] It offers a range of commercial services (home shopping, direct marketing, interactive advertising, coupon printing in the home, business "yellow pages," classified advertising, electronic catalogs), banking and financial services (including insurance), government services (e.g., home printing of fishing licenses in response to electronic application), other institutional information services (e.g., education, health), a form of "broadcast" electronic mail, and home-automation and energy-management services. Loto-Québec will provide lottery-related information and games, including interactive lotteries. The payment subsystem, managed by the National Bank of Canada, enables three forms of payment from the home (via credit card, via debit card, and via an electronic wallet embedded in the UBI smart card). The National Bank's plans to extend the use of the electronic wallet outside the home is indicative of the consortium members' efforts to capitalize on the network externalities and first-comer advantages implicit in a universal meta-audience.[62]

To succeed, UBI has to attract subscribers and ensure that they interact and transact with vendors. Even though the service is free, 80 percent of households have to allow the equipment (set-top box, remote control, printer, and payment terminal) to be installed and sign up for the smart card and the associated financial obligations. In the launch phase, the primary task is that of attracting subscribers and getting them to allocate non-negligible amounts of attention to the offered services. Pierre Dion (then general manager of marketing at Vidéoway Multimedia) put it this way:

Technology is one thing, but the real challenge is proving the concept. Do 80 percent of the people aged 7 to 70 want all these two-way services and will the product meet the hopes and ambitions of the partners, service providers and advertisers? If the answer is yes, then UBI can change the way people live.[63]

In subscribing to a service such as UBI, individuals choose to allocate part of their scarce attention to this service and not to others. As subscribers, they can allocate their attention to one or more of the vendors, watch television channels provided over the same facility, or do something not involving the UBI network. In other words, UBI's challenges are to induce individuals and households to join the meta-audience known as the UBI network and to create conditions for subscribers' attention to be allocated to vendors through the assembly of meso-audiences and other incentives. Vendors must assemble their audiences using the various functionalities provided by UBI. The economic viability of the network depends on the iteration of these acts.

UBI's designers have two sets of information-collection incentives. The first set is conducive to trust and privacy. Because its revenues are a function of the use of information services, the required information is the volume of traffic from the network to the vendors, not the individual users' TGI. The second set consists of incentives to collect, analyze, and sell information about subscribers to UBI vendors and others. That is, a vendor may be sold information on which subscribers use its services for what duration (i.e., individual profiles), or a current or potential vendor may be provided with information on the overall use patterns of subscribers (i.e., aggregate data). For the most part, the design of UBI ignores the second set of incentives. Indeed, this system appears to have more privacy safeguards built into it than many (perhaps any) of the interactive telecommunication projects that preceded it in the United States, Canada, and Europe. The promotional literature emphasizes privacy safeguards:

UBI will not keep any client files and will not bill the consumer; service suppliers, along with cable operators, will be billed and will make up the main source of income for financing the infrastructure.[64]

Furthermore, the consortium has developed a comprehensive code of conduct binding on the consortium, the partners, and all service providers. The code, developed with extensive public consultation, is incorporated into contracts between UBI and the vendors and includes enforcement procedures and an ombudsman.[65]

UBI's design exemplifies the trust-conducive structures of interactive systems producing universal meta-audiences. UBI's viability depends on attracting 80 percent of households in a market and creating conditions for a reasonable amount of subscribers' attention to be allocated to the revenue-generating vendors through the assembly of meso-audiences and other incentives. Teletel/Minitel (the French videotex initiative) created its meta-audience by cross-subsidizing the network enhancements from its monopoly revenues, giving free terminals and monthly service to households (again, enabled by its franchised monopoly status), and substituting an online directory accessible through Teletel for the print directory.[66]

The UBI design follows the "carrot" aspects of the Teletel/Minitel solution: network enhancements are paid for by a consortium of firms, each dominant in its market and some exclusively franchised, and the terminal and monthly service are provided free to households. It does not utilize the "stick" aspect of migrating a currently provided, essential service to the interactive network. It also does not enjoy a monopoly environment such as France Télécom enjoyed when it established Teletel/Minitel. UBI has a pack of competitors at its heels, including deep-pocketed Canadian telephone companies currently rolling out their own interactive, broadband Beacon/Sirius network in the neighboring province of New Brunswick.[67] Therefore, UBI has to "sweet-talk" the people of its first market, the Saguenay region, to join the system.

All or most of the households cannot be cajoled to subscribe by free terminals and service alone; the active interest and support of an overwhelming majority of the populace must be won. A firm in this unusual situation has to accommodate minorities, including "privacy fundamentalists."[68] If even a few opinion leaders become disaffected, if even a few opponents gain media attention, or if the media turn hostile, the battle is lost. The process resembles a political campaign where an 80 percent vote is required to win, rather than a marketing campaign. UBI cannot respond to high-surveillance incentives for fear of triggering opposition from Québec's active and articulate consumer-rights and citizen-rights groups. Any opposition would scuttle the 80 percent yes vote.[69]

According to UBI's planners, the decision to create an enforceable code binding on consortium members and vendors arose from research conducted for two separate business plans.[70] The business plans and the research are proprietary, but other survey results are instructive. The recent survey by Louis Harris Associates for the newsletter *Privacy and American Business* probed potential users' interest in and concerns about customer profiles, an element of many marketing strategies for interactive services. A key finding was that the stated take-up rate for interactive services utilizing customer profiles would double with privacy safeguards.[71]

The consortium commissioned the Public Law Research Center at the Université de Montréal to develop a comprehensive code of conduct binding on itself, on the partners, and on all vendors.[72] In the quotation used by UBI for the press release, Pierre Trudel, the Director of the Center, states:

Our approach is to protect the individual rights and preserve the values society holds dear by means of a *preventive, proactive approach* rather than a purely defensive strategy. By establishing a code of ethics that addresses the concerns and values of *all* users in a network such as UBI, the company can go a long way toward ensuring that new services will be introduced with a minimum of difficulty.[73]

The code is incorporated into vendor contracts and includes enforcement procedures. A consultative process that included more than 30 public meetings in the Saguenay region (the site of phase I) and in Montréal elicited about 40 written comments.[74] The final document, comprising 142 articles, refers to and incorporates relevant Québec and federal laws, and includes stricter provisions in some cases (for example, limiting the creation of individual consumer profiles by the consortium).[75]

Referring to the roles of the various partners in the consortium, the outside legal expert reported that, although there had been "hard discussion," no partner was "completely reluctant" regarding privacy. In his opinion, different corporate cultures affected their positions, in particular the cultures of the government-owned partners (Canada Post, Hydro-Québec, and Loto-Québec). These partners were highly sensitive to political considerations. However, even Hearst, the American partner, had been committed to privacy.[76] Canada Post, which describes itself as an "honest broker," sees privacy as an integral element of its traditional service that has to be carried over to the virtual world. Though privacy could have been built into the postal subsystem of the network only, Canada Post's and UBI's marketing research showed that people feared "new ways of doing things." As a result, Canada Post promoted the code of conduct as a means of making the users comfortable and of reaching the desired penetration rate.[77]

The Canada Post representative's comments on people's fear of new ways and levels of comfort were echoed by almost all the interviewees associated with the design process. UBI had taken a major step in making use easy by incorporating a sophisticated payment mechanism that allowed multiple payment modes, the ability to make small payments, the ability to transfer funds from home (e.g., from a bank account to an electronic wallet), receipt printing, and so on. Beyond that, the designers still wanted to create a "secure and comfortable environment" for users. The code and the high level of consultation and publicity given to it appear to be integral to this effort. By directly addressing the concerns identified in business-plan studies, surveys, and various consultative meetings, the UBI consortium appears to be creating a discourse intended to make users secure and comfortable in the virtual space of the network.

The UBI consortium makes no pretense that it is engaged in some form of public service or that it is safeguarding privacy for its own sake. The primary focus of the consortium is profit (and perhaps the gaining of first-comer advantages that are considerable in networked systems[78]). Its commercial interests are better served by renouncing the revenue and control potential of interactive technology. The low-surveillance design and the associated discursive processes offer better prospects of building the requisite universal meta-audience and effective audience assembly by vendors.

The UBI Code

A low-surveillance design will not build trust by itself. Subscribers must know that the system is trustworthy. The highly publicized UBI Code of Conduct plays a crucial role in this regard. The code assures users that information collection by the network is minimal, that information not required for payment and operational purposes will not be retained, and that safeguards apply to collected information. The payment agent provides even less information to vendors than a bank handling credit-card and debit-card payments in conventional, proximate settings.[79] Vendors may collect information (subject to limitations described below) from subscribers who use their services, but are prohibited from selling or transferring such information to third parties.

The first part of the code affirms the rights of users to a convivial electronic environment respectful of the values of dignity and honesty that form the basis of exchange and commerce. This is followed by a section on personal information protection and privacy, which is basically a transposition of relevant Québec legislation, particularly the 1993 Act Respecting the Protection of Personal Information in the Private Sector (hereafter referred to as the Private Sector Privacy Act) and the applicable federal privacy provisions.[80]

The code sets out obligations pertaining to the smart card, proof of transactions, and security of transactions. The obligations of service providers include an appropriate level of security (high for financial transactions, low for electronic flyers). The code creates a uniform regime for electronic shopping by incorporating and clarifying the various provisions in Québec's consumer law for door-to-door, mail-based, and telephone-based sales. It assumes that every vendor will have a business presence in Québec and will be bound by Québec and federal law. The problem of service providers without business presence in the province, acknowledged to be a difficult one, had been postponed for future action. The code includes sections on financial services in the home, electronic mail services, energy-management services, and the content of information services (the last covering issues such as violence and advertising directed to children).

Finally, the code sets out alternative dispute-resolution procedures and makes an internal committee the last level of appeal within the organization. Consumers have access to normal legal procedures whether or not they use the internal mechanisms. The internal committee is mandated to harmonize internal procedures with applicable provincial and federal laws.

The interviews identified areas of ambiguity in the code, which in fact also exist in the Québec legislation and among stakeholders.[81] The first was the definition of the phrase "information necessary for the object of the file" (referring to the file created and maintained by the business enterprise), found in articles 4 and 5 of the Private Sector Privacy Act. The code does not define "information necessary for the object" of a UBI transaction. For example, electronic-wallet transaction records are arguably not necessary for the provision of "electronic cash." Were UBI 's payment agent to refrain from collecting or retaining such information, at least one of the three payment options would offer true anonymity.

However, the National Bank intends to collect and retain such records for up to 6 months, defining them as "necessary."[82] The outside expert recognized the ambiguity and voiced hope that practice and judicial interpretation of a relatively new law would provide the needed clarifications. The code includes specific prohibitions on service providers' making the offer of goods or services conditional on the granting of permission to collect, use, or disseminate personal information beyond what is "absolutely necessary for the transaction." This follows Article 9 of the Private Sector Privacy Act:

No person may refuse to respond to a request for goods or services . . . by reason of the applicant's refusal to disclose personal information except where
(1) collection of information is necessary for the conclusion or performance of a contract;
(2) collection of that information is authorized by law; or
(3) there are reasonable grounds to believe that the request is not lawful.

In case of doubt, personal information is considered not necessary. Although this is an advance over much of the existing privacy law in other jurisdictions, the value of these provisions depends on what information is deemed necessary for a transaction.[83]

The second area of ambiguity affects the transfer of information collected and retained by the UBI network or by subsystem administrators within the organizations (e.g., transfer of payment-subsystem information to another part of the National Bank for the marketing of financial services). This is an issue common to all interactive platforms, including the public telecommunication network.[84] UBI representatives claimed that such transfers would not occur, since the following provisions of the Private Sector Privacy Act had been included in the code:

Article 13: No person may communicate to a third person the personal information contained in the file he holds on another person, or use it for purposes not relevant to the object of the file, unless the person concerned consents thereto or such communication or use is provided for by this Act.

Article 14: Consent to the communication or use of personal information must be manifest, free, and enlightened, and must be given for specific purposes. Such consent is valid only for the length of time needed to achieve the purposes for which it was requested. Consent given otherwise than in accordance with the first paragraph is without effect.

In addition, such transfers were claimed to be precluded by the segmented design and, in the case of the payment agent, by the bank's internal security and privacy policies.[85] It was conceded, however, that transfers were technically feasible—particularly in the case of the bank, which kept records of all transactions for up to 6 months.[86] The legality of various forms of information sharing depends on the construction of "third person" and "purposes not relevant to the objects of the file." If a third person is defined as an entity with a different legal personality, transfers within an organization such as the National Bank would not be prohibited. The ambiguity of the second phrase was discussed above.

The above discussion of the privacy safeguards built into the overall UBI system, particularly through the code of conduct binding on all consortium members and vendors, demonstrates the significance of the legal environment of Québec. The civil-law legal culture of this fiercely independent province differs from that found in the common-law jurisdictions of English Canada and the United States. The privacy-protection regime in Québec comprises a set of related documents, including the Québec Charter of Human Rights and Freedoms (hereafter, Québec Charter), the Civil Code of Québec (hereafter, Civil Code), the act respecting access to documents held by public bodies and the protection of personal information (hereafter Public Sector Privacy Law), and the 1993 Private Sector Privacy Law. These laws, taken together, constitute the most advanced privacy-protection regime in North America.[87]

The province's fundamental law, the 1975 Québec Charter, was adopted during the first wave of privacy legislation in industrialized market economies.[88] Reflecting the tenor of the times, it assures, among others, the following rights:

5. Every person has a right to respect for his private life. . . .

7. A person's home is inviolable. . . .

9. Every person has a right to non-disclosure of confidential information (Québec, L.R.Q., c. C-12).

The other basic law, the Civil Code has a chapter entitled "Respect of reputation and privacy." The Civil Code, adopted in 1987, includes the following among its technology-neutral provisions on privacy:

37. Every person who establishes a file on another person shall have a serious and legitimate reason for doing so. He may gather only information

which is relevant to the stated objective of the file, and may not, without the consent of the person concerned . . . communicate such information to third persons or use it for purposes that are inconsistent with the purposes for which the file was established.

The Public Sector Privacy Act (1982) and the Private Sector Privacy Act (1993) apply these abstract principles of privacy to concrete situations and establish enforcement mechanisms. As has been shown with regard to some of the sections incorporated into the UBI code, these provisions are both general in effect (in contrast to the United States' sectoral approach) and specific with regard to what is allowed and not allowed.

Conclusions

Dystopian prognostications pertaining to surveillance and interactive networks have been based on one-sided analyses. The underlying economic processes and the accompanying processes of channel proliferation and audience fragmentation hold the potential for consensual as well as coercive surveillance. Customized mass production makes customer relationships necessary, but it does not predetermine whether they will be stable, productive relationships based on trust, privacy, and consensual surveillance or whether they will be pathological relationships based on coercive surveillance, mistrust, and angst.

Interactive media systems are called forth by trends in the economy toward mass customization. Interactive systems can be used for overt or covert coercive surveillance (as is the case for the most part today) or for the creation of consensual surveillance. The scarcity of attention (currently exacerbated by the proliferation of messages and channels), when combined with the incentives of network externalities and first-comer advantages, makes the formation of a universal meta-audience an attractive business opportunity. A system operator creating a universal meta-audience based on an interactive media system has incentives to design a trust-conducive environment, and to cajole and even coerce vendors to cooperate in building and maintaining it. In view of the incentives all commercial organizations have to build and maintain stable, productive relationships with customers, and in view of the difficulties of engaging in covert coercive surveillance on an interactive system designed and controlled by a third party, vendors are likely to cooperate.

Institutional and strategic-conduct analysis of the UBI network in Québec reinforces the above theoretical conclusions. The analysis suggests that trust-conducive and privacy-conducive interactive systems are likely to emerge when all of the following necessary conditions are satisfied:

1. The system operator must seek to produce a universal meta-audience.
2. All or most of the system operator's revenues must be derived from vendors in return for access to the meta-audience, meso-audiences, and functionalities such as payment systems.
3. Privacy concerns must be present in the populace and must be known to corporate managers through proprietary market research, through publicly available privacy surveys, or through the existence of active privacy advocacy groups. The existence of privacy legislation may be seen as a proxy indicator of privacy concerns among the populace.[89]
4. Some affirmative acts on the part of the individual consumer must be required before he or she joins the interactive network.

These conditions apply to the time before the network becomes fully operational, and do not necessarily apply to a functioning network.

The necessary conditions rest to a large extent on questions about how much the system operators, the vendors, and the consumers know. System operators and vendors are likely to learn about the prerequisites of stable, productive customer relationships in the process of achieving universality in a meta-audience and inducing meta-audience members to use the system. However, this education may not occur in an environment that allows consumers to be "signed onto" interactive systems by default (without an affirmative act on their part). The "creeping" redesign of public telecommunication networks throughout the world to include covert surveillance capabilities and policies that do not give subscribers adequate information about the changes and the choice to accept or reject them exemplify violation of "knowledgeability conditions."[90] In this situation, not only do system operators and vendors lose the opportunity to learn about the trust and privacy concerns of the public; in addition, individual consumers lose the opportunity to enter into satisfying commercial relationships. The resulting dissonance may lead to mistrust and angst, resulting in pathological customer relationships.

The outcome of the current process wherein system operators, vendors, government, and interest groups are shaping the manner in which surveillance, privacy, and trust are addressed in the design of interactive media systems is not preordained. Outcomes that are good from the perspective of privacy advocates are possible, particularly if corporate actors learn the benefits of trust-based commercial relationships. However, outcomes that are bad from the perspective of privacy advocates may result, particularly if public knowledge of the complexities of the emerging interactive systems does not improve. Public attitudes are affected by laws and regulatory processes as well as by corporate practices. Abstract systems that are designed to engage in routine coercive surveillance can lead to changes in public attitudes that are now favorable to privacy. In the same way that the design of government and commercial systems has made people treat the yielding of social security numbers as a taken-for-granted act, routinization of coercive surveillance may change public attitudes toward privacy. Coercive surveillance will then not result in mistrust, angst, and pathological customer relationships. Incentives for system operators and vendors to safeguard privacy and create trust-conducive environments will weaken and disappear. The private and public policy actions taken during the current period, in which a great majority of the populace is concerned about coercive surveillance in an unfocused yet persistent way, may be decisive in terms of the path that will be taken. Once coercive surveillance becomes routinized and taken for granted, the prospects for privacy and trust conducive outcomes are likely to be quite dim.

Acknowledgements

The research for this chapter was supported by grants from the Canadian Studies Research Program of the Canadian Embassy in the United States and from the Ministère des affaires internationales, de l'immigration et des communautés culturelles of the Province of Québec. The author is grateful to Ashwini Tambe for research and translation assistance; to Stephanie Perrin and Marie Valeé for initial encouragement; to Pierrôt Péladeau for intellectual exchanges and information; to Huichuan Liu, Peter Shields, Phil Agre, Oscar Gandy, David Flaherty, and attendees at a presentation at the Office of the Information and Privacy Commissioner of British Columbia for comments; to the Ohio State University for a professional leave; and to all the busy individuals associated with UBI for making the time for interviews.

Notes

1. T. McManus first proposed this term in the report Telephone Transaction-Generated Information: Rights and Restrictions (Harvard University Center for Information Policy Research, 1990). TGI is a by-product of an interaction or a transaction.

2. M. Aglietta, *The Theory of Capitalist Regulation: The US Experience* (New Left Books, 1979); M. Castells, The informational economy and the new international division of labor, in *The New Global Economy in the Information Age*, ed. M. Carnoy et al. (Pennsylvania State University Press, 1993); P. Cooke and K. Morgan, The network paradigm: New departures in corporate and economic development, in *Environment and Planning D: Society and Space* 11 (1993): 543–564; D. Harvey, *The Condition of Post-Modernity: An Enquiry into the Origins of Cultural Change* (Blackwell, 1990); B. J. Pine II, *Mass Customization: The New Frontier in Business Competition* (Harvard Business School Press, 1993); M. J. Piore and C. F. Sabel, *The Second Industrial Divide: Possibilities for Prosperity* (Basic Books, 1984); R. B. Reich, *The Work of Nations: Preparing Ourselves for 21st Century Capitalism*, second edition (Vintage, 1992).

3. Pine, *Mass Customization*, p. 44. The term was originally proposed in S. M. Davis's *Future Perfect* (Addison-Wesley, 1987).

4. A. Bressand, C. Distler, and K. Nicolaïdis, Networks at the heart of the service economy, in *Strategic Trends in Services*, ed. A. Bressand and K. Nicolaïdis (Ballinger, 1989).

5. Pine, *Mass Customization*, p. 46.

6. M. J. Weiss, *The Clustering of America* (Harper & Row, 1988).

7. O. H. Gandy Jr., *The Panoptic Sort: A Political Economy of Personal Information* (Westview, 1993); E. Larson, *The Naked Consumer* (Holt, 1992). See also Pine, *Mass Customization*, p. 44, on the critical importance of understanding customers' requirements and on the role of TGI in the financial-services industry's move to the forefront of mass customization.

8. Bressand et al., Networks at the heart of the service economy. D. W. Smythe made a similar claim, independent of and before Bressand et al., in *Dependency Road: Communications, Capitalism, Consciousness, and Canada* (Ablex, 1981), p. xiv.

9. J. Rule, D. McAdam, L. Stearnes, and D. Uglow, *The Politics of Privacy: Planning for Personal Data Systems as Powerful Technologies* (Elsevier, 1990).

10. For an analysis of minimum information conditions for virtual and proximate forms of service delivery, see H. Liu, Privacy-Implicated System Design in the Virtual Marketplace, Ph.D. dissertation, Ohio State University, 1996.

11. R. Samarajiva, Privacy in electronic public space: Emerging issues, *Canadian Journal of Communication* 19 (1994), no. 1, p. 92.

12. David Shepard Associates, Inc., *The New Direct Marketing: How to Implement a Profit-Driven Database Marketing Strategy*, second edition (Dow Jones–Irwin, 1995).

13. No claim is made about actual media enterprises. Through concentration of ownership or alliances, a media enterprise may control multiple channels.

14. D. Peppers and M. Rogers, *The One-to-One Future: Building Relationships One Customer at a Time* (Doubleday, 1993).

15. V. Mosco, *The Pay-Per Society: Computers and Communication in the Information Age* (Ablex, 1989).

16. The concerns regarding cable-usage records that emerged in conjunction with QUBE (Warner Cable's first interactive trial in Columbus, Ohio) led to the first code of conduct for cable service, which was subsequently codified. See D. H. Flaherty, *Protecting Privacy in Two-Way Electronic Services* (Knowledge Industry Publications, 1985); C. Davidge, America's talk-back television experiment: QUBE, in *Wired Cities*, ed. W. Dutton et al. (G. K. Hall, 1987). US law provides greater protection for video-rental records than for health records; see R. Mukherjee and R. Samarajiva, The customer web: Transaction generated information and telecommunication, *Media Information Australia* 67 (1993), February, p. 51.

17. In an oft-quoted dissent, Justice William O. Douglas wrote that the checks a person wrote could disclose "a fairly accurate account of his religion, ideology, opinions, and interests" (*California Bankers Association v. Schultz*, 416 US 21 (1974)).

18. M. Foucault, *Discipline and Punish: The Birth of the Prison* (Vintage, 1977), p. 191.

19. See, e.g., E. I. Schwartz, People are supposed to pay for this stuff? *Wired* 3.07 (1995): 149–153, 187–191.

20. S. G. Steinberg, Software metering, *Wired* 3.07 (1995): 137–138.

21. P. Samuelson, The copyright grab, *Wired* 4.01 (1996): 135–138, 188–191.

22. K. J. Arrow, Gifts and exchanges, in *Altruism, Morality and Economic Theory*, ed. E. Phelps (Russell Sage Foundation, 1975), p. 24.

23. A. Giddens, *The Consequences of Modernity* (Stanford University Press, 1990), p. 121.

24. Ibid., pp. 99–100.

25. Ibid., p. 34.

26. Ibid., p. 33.

27. Ibid., pp. 99–100.

28. D. Boden and H. L. Molotch, The compulsion of proximity, in *NowHere*, ed. R. Friedland and D. Boden (University of California Press, 1994); Giddens, *Consequences of Modernity*, pp. 83–111.

29. Giddens, *Consequences of Modernity*, p. 121.

30. E. Goffman, *Relations in Public: Microstudies of the Public Order* (Basic Books, 1971), p. 198.

31. I. Altman, *The Environment and Social Behavior: Privacy, Personal Space, Territory, and Crowding* (Books/Cole, 1975); S. Petronio, Communication boundary management: A theoretical model of managing disclosure of private information between marital couples, *Communication Theory* 1 (1991), no. 4: 311–335; M. Ruggles, Mixed signals: Personal data control in the intelligent network, *Media Information Australia* 67 (1993), February: 28–39.

32. Samarajiva, Privacy in electronic public space, p. 90.

33. S. D. Warren and L. D. Brandeis, The right to privacy, *Harvard Law Review*, 4 (1890), drawing from T. M. Cooley, *A Treatise on the Law of Torts or the Wrongs Which Are Independent of Contract* (Callaghan & Co., 1879).

34. This was aptly expressed in 1856 in relation to asylums: "It is essential in all intercourse with the patients that the attendant's conduct should be soothing. It must never be distrustful; but above all, while a constant surveillance is necessary, it is important that he be not obtrusive or unnecessarily interfering."—W. H. O. Sankey, Do the public asylums of England, as at present constructed, afford the greatest facilities for the care and treatment of the insane? *Asylum Journal of Mental Science* 2 (1856): 470, quoted in C. Philo, "Enough to drive one mad": The organization of space in 19th-century lunatic asylums, in *The Power of Geography*, ed. J. Wolch and M. Dear (Unwin Hyman, 1989).

35. Giddens, *Consequences of Modernity*, pp. 84–85.

36. Co-present interactions with system representatives tend to be limited to signing-up and installation. These interactions clearly fit Giddens's discussion. Subsequent interactions tend to not involve co-presence, generally occurring via some technological medium such as the telephone, and they tend to occur in situations where the routine has been disrupted. Increasingly, telephone, cable, and energy companies are dispensing with co-present interactions even at the stage of signing up and installation.

37. Fédération nationale des associations de consommateurs du Québec and Public Interest Advocacy Centre, *Surveying Boundaries: Canadians and Their Personal Information* (Public Interest Advocacy Centre, 1995); Louis Harris & Associates, Interactive Services, Consumers and Privacy (Privacy & American Business, 1994).

38. Gandy, *Panoptic sort*. On general corporate information practices, see Larson, *Naked consumer*, and H. J. Smith, *Managing Privacy: Information Technology and Corporate America* (University of North Carolina Press, 1994).

39. M. Csikszentmihalyi, *Flow: The Psychology of Optimal Experience* (Harper & Row, 1990), pp. 28–33.

40. Quoted in H. R. Varian, The information economy: How much will two bits be worth in the digital marketplace? *Scientific American*, September 1995: 200–202. See also W. R. Neuman, *The Future of the Mass Audience* (Cambridge University Press, 1991), p. 150.

41. Persuasion is seen by some to be coercive. The term is used differently here. Some forms of persuasion may be coercive, but not all forms of persuasion necessarily are. It is difficult to clearly distinguish between communication that is persuasive and communication that is not.

42. *Oxford English Dictionary* (1989), "market," definition 1(a): "The meeting or congregating together of people for the purchase and sale of provisions or livestock, publicly exposed, at a fixed time and place; the occasion, or time during which such goods are exposed to sale; also, the company of people at such a meeting."

43. This chapter focuses on the production and reproduction of audiences for political and economic organizations. Interpersonal and small-group interactions also require attention. These other forms of allocating attention are not dealt with here.

44. *Oxford English Dictionary* (1989), "audience," meanings, 7(b) and 7(c). The audience for a television or radio program suggests concentration of attention at a given time; the audience for a book suggests that simultaneity is not essential.

45. For overviews see I. Ang, *Desperately Seeking the Audience* (Routledge, 1991); L. Jeffrey, Rethinking audiences for cultural industries: Implications for Canadian research, *Canadian Journal of Communication* 19 (1994): 495–522; Neuman, *Future of the Mass Audience*.

46. W. H. Melody, *Children's Television: The Economics of Exploitation* (Yale University Press, 1973); D. W. Smythe, Communications: Blindspot of Western Marxism, *Journal of Political and Social Theory* 1 (1977), no. 3: 1–27. See also B. M. Owen, J. H. Beebe, and W. G. Manning Jr., *Television Economics* (Lexington Books, 1974); B. M. Owen and S. S. Wildman, *Video Economics* (Harvard University Press, 1992). For industry perspectives, see D. Poltrack, *Television Marketing: Network, Local, and Cable* (McGraw-Hill, 1983); H. L. Vogel, *Entertainment Industry Economics: A Guide for Financial Analysis*, third edition (Cambridge University Press, 1994).

47. Ang, *Desperately Seeking the Audience*. For earlier mainstream social science work on these lines, see R. A. Bauer, The audience, in *Handbook of communication*, ed. I. de Sola Pool et al. (Rand McNally, 1973).

48. J. Meyrowitz, *No Sense of Place: The Impact of Electronic Media on Social Behavior* (Oxford University Press, 1985); J. Meyrowitz, Medium theory, in *Communication Theory Today*, ed. D. Crowley and D. Mitchell (Stanford University Press, 1994). This tradition of scholarship originated with H. A. Innis (*The Bias of Communication* (University of Toronto Press, 1951/1964)) and M. McLuhan (*The Gutenberg Galaxy: The Making of Typographic Man* (University of Toronto Press, 1962).)

49. *Oxford English Dictionary* (1989), "audience," meaning 7(b).

50. J. Turow, On reconceptualizing "mass communication," *Journal of Broadcasting and Electronic Media* 36 (1992), no. 1: 105–110. By giving prior-

ity to the industrial nature of the production, reproduction, and multiple distribution of messages, Turow attempts to resuscitate the concept "mass communication," originally defined in relational terms by C. R. Wright (*Mass Communication: A Sociological Perspective* (Random House, 1959)). However, Turow recognizes that the potential mass audience created by these industrial processes may rarely, if ever, be realized. If the mass audience rarely, if ever, materializes, the "mass" appellation appears to have little value.

51. Dayparts "are not strictly defined by a time specification; they represent program groupings reflecting similar audience composition" (Poltrack, *Television marketing*, p. 49).

52. M. W. Miller, Lobbying campaign, AT&T directories raise fears about use of phone records, *Wall Street Journal*, December 13, 1991.

53. Vogel begins his discussion of the economics of broadcasting (*Entertainment Industry Economics*, p. 152) with the aphorism "Programs are scheduled interruptions of marketing bulletins." Smythe described programs, after A. J. Liebling, as the "free lunch" intended to attract viewers to advertisements (*Dependency Road*, p. 37).

54. The term is adapted from the concept of universal service in telephony.

55. The delays and problems in the startup of UBI reported by J.-H. Roy (La mort d'une inforoute? UBI titube, *Voir*, September 27, 1995), and even UBI's possible failure, will not affect the thesis developed in this chapter.

56. W. B. Arthur, *Increasing Returns and Path Dependence in the Economy* (University of Michigan Press, 1994).

57. K. Dougherty, Groupe Vidéotron set to offer home services through coaxial cable, *Financial Post*, December 24, 1994 (NEXIS database).

58. Université de Montréal, Centre de recherche en droit public, Code de conduite proposé pour l'environnment UBI, Document de travail, November 8, 1995. The set-top box and the associated terminal devices are likely to be free of charge, though some deposits may be required.

59. R. Thivierge, interview, July 11, 1995, UBI, Montréal.

60. R. W. Hough, Alternative routes to telco broadband in local markets, presented at Telecommunications Policy Research Conference, Solomons, Maryland, 1995; J. Markoff, Phone companies hit interactive-TV snags, *New York Times*, October 1, 1994.

61. Interviews showed this to be the dominant metaphor among the designers.

62. S. Rataczack, interview, July 12, 1995, National Bank of Canada, Montréal.

63. R. Gibbens, Vidéotron aims to connect with interactive TV test, *Financial Post -Weekly*, October 15, 1994.

64. UBI, Basic Information Guide, November 1994.

65. Université de Montréal, Centre de recherche en droit public, Code de conduite; P. Trudel, interview, July 13, 1995, Université de Montréal.

66. *European Telematics: The Emerging Economy of Words*, ed. J. Jouët et al. (North-Holland, 1991); J.-M. Charon, Videotex: From interaction to communication, *Media, Culture and Society* 9 (1987): 301–332.

67. G. Van Koughnett, Présentation: Réglementation, déreglementation et autoroutes électroniques, in *Les autoroutes de l'information un produit de la convergence*, ed. J.-G. Laclroix and G. Tremblay (Presses de l'université Québec, 1995).

68. A. F. Westin, Privacy rights and responsibilities in the next era of the information age, in *Toward an Information Bill of Rights and Responsibilities*, ed. C. Firestone and J. Schement (Aspen Institute, 1995), p. 86. Westin's methodology is well established and influential, though not uncontested. Westin claims that about 25% of the US populace are privacy fundamentalists.

69. The general problem is analyzed in A. Hirschman, *Exit, Voice, and Loyalty* (Harvard University Press, 1970).

70. R. Thivierge, interview, July 29, 1994, UBI, Montréal.

71. Louis Harris and Associates, *Interactive Services, Consumers and Privacy*, pp. xvi–xvii.

72. UBI, "The UBI consortium gives itself a code of ethics, to be drafted by the University of Montréal Centre de recherché en droit public," Press release distributed on PR Newswire, June 21, 1994; Trudel, interview.

73. UBI, "The UBI consortium gives itself a code of ethics" (emphases added).

74. For example, in November 1994 a 44-page document entitled "MEMOIRE presente dans le cadre de la consultation en vue de l'adoption d'un code de déontologie pour UBI" was presented on behalf of the Institut canadien d'education aux adultes (a specialized organization working in the field of adult education, public broadcasting, and communication policy on behalf of a huge network of community groups and trade unions), the Federation nationale des communications (a federation of trade unions representing workers in newspapers, broadcasting, and other sectors, including a trade union of Vidéotron workers), and the Confederation des syndicats nationaux (the confederation of Quebec national trade unions) (P. Péladeau, personal communication, September 22, 1995).

75. Trudel, interview.

76. Ibid.

77. M. Labrèque, interview, July 12, 1995, UBI, Montréal.

78. Arthur, *Increasing Returns and Path Dependence*. The competitive implications of achieving 80% penetration for interactive access and for the electronic wallet (the use of which is not limited to the UBI network) are significant.

79. Rataczack, interview.

80. This (described more fully below) is the only statute imposing general privacy-protection obligations on the private sector in North America.

81. For example, the Fédération nationale des associations de consommateurs du Québec (FNACQ), a major consumer-rights organization with a strong

record of privacy advocacy, believes that the collection and retention of consumer transaction information is desirable for the purpose of resolving disputes.

82. Rataczack, interview.

83. See extensive discussion in O. H. Gandy Jr., Legitimate business interest: No end in sight, *Chicago Legal Forum* (forthcoming).

84. R. E. Burns, R. Samarajiva, and R. Mukherjee, Utility Customer Information: Privacy and Competitive Implications, Report 92-11, National Regulatory Research Institute, Columbus, Ohio, 1992.

85. Rataczack, interview; Trudel, interview.

86. Rataczack, interview.

87. N.-A. Boyer, The road to legislation, part 1: The story behind Quebec's Bill 68, *Privacy Files* 1 (1996), no. 5: 7–8; P. Péladeau, Human rights and GII: From data protection to democratization of infrastructure decisions, presented at INET'96 Conference, Montréal.

88. C. J. Bennett, *Regulating Privacy: Data Protection and Public Policy in Europe and the United States* (Cornell University Press, 1992).

89. P. Péladeau, interview. July 10, 1995, FNACQ office, Montréal.

90. R. Samarajiva, Surveillance by design: Public networks and the control of consumption, in *Communication by Design,* ed. R. Mansell and R. Silverstone (Oxford University Press, 1996).

Contributors

Philip E. Agre
Department of Communication
University of California, San Diego
La Jolla, CA 92093-0503
pagre@ucsd.edu
http://communication.ucsd.edu/pagre/

Victoria Bellotti
Apple Computer, Inc.
1 Infinite Loop, MS: 301-4A
Cupertino, CA 94043
bellotti@apple.com

Colin J. Bennett
Department of Political Science
University of Victoria
Victoria V8W 3P5, BC
cjb@uvic.ca

Herbert Burkert
GMD - German National Research Center for Information Technology
Schloß Birlinghoven
D-53754 Sankt Augustin, Germany
burkert@gmd.de
http://www.gmd.de/People/Herbert.Burkert/

Simon G. Davies
Privacy International
666 Pennsylvania Ave SE, Suite 301
Washington, DC 20003
davies@privint.demon.co.uk
http://www.privacy.org/pi/

David H. Flaherty
Information and Privacy Commissioner for British Columbia
1675 Douglas Street
Victoria V8V 1X4, BC
dflaherty@galaxy.gov.bc.ca
http://www.cafe.net/gvc/foi

Robert Gellman
Privacy and Information Policy Consultant
431 Fifth Street SE
Washington, DC 20003
rgellman@cais.com

Viktor Mayer-Schönberger
Rüdigergasse 19/15
A-1050 Wien, Austria
vms@gii.priv.at

David J. Phillips
Annenberg School for Communication
University of Pennsylvania
Philadelphia, PA 19104-6220
djp@pobox.asc.upenn.edu / djphill@umich.edu

Marc Rotenberg
Electronic Privacy Information Center
666 Pennsylvania Ave. SE, Suite 301
Washington, DC 20003
rotenberg@epic.org
http://www.epic.org

Rohan Samarajiva
School of Communication and Journalism
3016 Derby Hall
Ohio State University
Columbus, OH 43210-1339
rohan+@osu.edu
http://communication.sbs.ohio-state.edu/commdept/faculty.html

Name Index

Adleman, Leonard, 254–255
Aglietta, Michel, 303
Agre, Philip, 3, 14, 18, 22, 32, 39, 48–49, 121
Allen, Anita, 1
Altman, Irwin, 305
Ang, Ien, 306
Arrow, Kenneth, 282, 304
Arthur, W. Brian, 307–308
Athanasiou, Tom, 275

Bamford, James, 255–256
Barnes, Ralph, 34, 36–37
Bauer, Raymond, 306
Beebe, Jack, 306
Bellotti, Victoria, 4, 6, 8–9, 11, 67, 69, 77
Benedikt, Michael, 47
Beniger, James, 45
Bennett, Colin, 2–3, 5–6, 18–19, 21–22, 25, 52, 120–123, 141, 216, 220, 238, 309
Berg, Marc, 48
Bertino, Elisa, 17
Bijker, Wiebe, 18, 56, 248
Blaze, Matt, 264
Bly, Sara, 66, 72, 76
Boal, Iain, 164
Boden, Dierdre, 304
Borgida, Alexander, 42
Bowers, C., 49
Bowers, John, 48
Bowker, Geoffrey, 16, 18

Bowyer, Kevin, 65
Boyer, Nicole-Ann, 309
Brachman, Ronald, 42
Brackett, Michael, 16
Brandeis, Louis, 141, 193, 202–203, 217, 305
Branscomb, Lewis, 3
Branstad, Dennis, 65
Braverman, Harry, 33
Brennan, William, 204
Brenton, Myron, 2
Bressand, Albert, 278–279, 303
Brook, James, 164
Brooks, John, 163
Bruce, James, 265
Bud-Frierman, Lisa, 48
Bull, Hans Peter, 178–179, 227, 239
Burke, Sandra, 237
Burkert, Herbert, 4, 6, 9–11, 22, 52, 141–142, 241
Burns, Robert, 309
Burnham, David, 1
Buxton, William, 76

Campbell, Duncan, 162
Cantwell, Maria, 264
Casson, Mark, 13–15, 20
Castells, Manuel, 303
Cavoukian, Ann, 25, 123
Chaum, David, 23, 52, 65, 268
Chen, Peter Pin-Shan, 39
Charon, Jean-Marie, 308
Clarke, Roger, 3, 7, 54, 64, 141

Clement, Andrew, 31
Coase, Ronald, 13
Comeau, Paul-Andre, 122
Connor, Steve, 162
Cooke, Philip, 303
Cool, Colleen, 69
Cooley, Thomas, 305
Cotrell, Lance, 275
Couger, J. Daniel, 35, 38–39
Cromwell, Larry, 158
Cronon, William, 16
Csikszentmihalyi, Mihaly, 305

Dam, Kenneth, 273
David, Paul, 21
Davidge, Carol, 304
Davies, Simon, 6, 11, 15, 19–20, 162–164
Davis, Stanley, 303
Denning, Dorothy, 65
Derrida, Jacques, 35
Di Genova, Frank, 30
Dion, Pierre, 292
Diffie, Whitfield, 255–256
Distler, Catherine, 303
Dougherty, K., 307
Douglas, William, 202, 304
Dourish, Paul, 49, 64, 66, 69–73, 76
Drake, William, 19
Dunlop, Charles, 64

Edwards, Paul, 32, 47
Eldridge, Margery, 66, 88, 93
Elgesem, Dag, 141
Ernst, Morris, 2
Evan, William, 165

Feenberg, Andrew, 31
Fiat, Amos, 52
Fish, Robert, 65, 69
Flaherty, David, 2–3, 5, 20–21, 52, 120, 162, 164–165, 190, 209, 218, 237, 304
Flusty, Steven, 163
Forester, Tom, 68, 80
Foucault, Michel, 304

Freeh, Louis, 267
Freeman, R. Edward, 165
Freese, Jan, 157, 178
Friedman, Andrew, 32
Friedman, Batya, 49
Froomkin, A. Michael, 259

Gandy, Oscar H. Jr., 3, 18, 303, 305, 309
Garfinkel, Simson, 261, 265, 275
Gaver, William, 66, 69–70, 73–74
Gelernter, David, 42–49
Gellman, Robert, 3, 5, 123, 216
Gibbens, Robert, 307
Giddens, Anthony, 282, 285, 304–305
Gilbreth, Frank, 33, 35–36
Goffman, Erving, 8, 63, 74, 283, 305
Gore, Albert, 264
Gostin, Lawrence, 2
Grace, John, 178
Graham, Stephen, 163
Gregory, William, 40–41
Greenleaf, Graham, 121–123
Guillén, Mauro, 33

Halpert, Julie, 30
Harlan, John, 203
Harper, Richard, 75
Harrison, Steven, 64, 69, 73
Harvey, David, 303
Heath, Christian, 65, 70, 74–75
Heery, Dan, 163
Hellman, Martin, 255–256
Henke, David, 165
Hindus, Debora, 66
Hirschman, Albert, 308
Holvast, Jan, 157, 164
Honess, Terry, 163
Hood, Christopher, 120
Hough, Roger, 307
Hustinx, Peter, 123
Hutton, Will, 148, 162

Inness, Julie, 11
Innis, Harold, 306

Jeffrey, Lisa, 306
Jouët, Josiane, 308
Kallman, Ernest, 237
Karst, Kenneth, 240
Keller, James, 3
Kling, Rob, 64
Kosten, Freddy, 121–122
Kriele, Martin, 240
Kuitenbrouwer, Frank, 241

Labrèque, Michel, 308
Lamming, Michael, 66
Lampson, Butler, 65
Landau, Susan, 257, 275
Larson, Erik, 303, 305
Lasch, Christopher, 164
Latour, Bruno, 16, 248
Law, John, 56, 248
Levy, Steven, 141, 264
Lewis, Peter, 267
Liebling, A. J., 307
Lin, Herbert, 273
Liu, Huichuan, 303
Louie, Gifford, 75–76
Lovering, John, 163
Luff, Paul, 65, 70, 74–75
Luhmann, Niklas, 272
Lyon, David, 1, 31, 142

Mackay, Hugh, 149, 163
Mackay, Wendy, 77
MacLean, Allan, 77, 79
Macomber, S. Kingsley, 30
Major, John, 150
Manning, Willard G. Jr., 306
Mantei, Marilyn, 69
March, James, 49
Markoff, John, 273, 307
Maxeiner, James, 218
May, Andrew, 151
Mayer-Schönberger, Viktor, 2–3, 5–6, 10, 237
McAdam, Douglas, 303
McCarthy, John, 74
McClurg, Andrew, 67
McGuffin, Lola, 75

McKenna, Regis, 141
McLuhan, Marshall, 306
McManus, Thomas, 303
Melody, William, 306
Meyrowitz, Joshua, 306
Michelman, Frank, 240
Mierzwinski, Ed, 202
Milberg, Sandra, 67, 237
Miller, Michael, 307
Minneman, Scott, 66, 76
Molotch, Harvey, 304
Montgomery, David, 33
Monti, Mario, 121
Moran, Thomas, 76
Morgan, Kevin, 303
Morrison, Perry, 68, 80
Morrow, Lance, 148, 162
Mosco, Vincent, 281, 304
Mukherjee, Roopali, 304, 309
Mullender, Sape, 65

Nader, Ralph, 144
Naor, Moni, 52
Negrey, Cynthia, 162
Nelson, Daniel, 33
Neuman, W. Russell, 305
Newman, William, 66
Nicolaïdis, Kalypso, 19, 303
Nissenbaum, Helen, 49
Nugter, Adriana, 2

Olsen, Johan, 49
Olson, Gary, 75
Owen, Bruce, 306
Ozsu, M. Tamer, 17

Packard, Vance, 2
Parent, William, 191
Parker, Donn, 64
Peck, Shaun, 183
Pedersen, Elin, 65
Péladeau, Pierrôt, 308–309
Peppers, Don, 304
Peritt, Henry, 141
Pessoa, Fernando, 141

Petronio, Sandra, 305
Phillips, David, 4, 6, 9–11, 19–20, 23, 57
Pierce, Kenneth, 255–256, 275
Pine, B. Joseph II, 303
Piore, Michael, 303
Pipe, G. Russel, 241
Pocock, John, 240
Podlech, Adalbert, 223, 238
Poltrack, David, 306–307
Pool, Ithiel de Sola, 31, 121
Porter, Theodore, 16
Portillo, Michael, 164
Posner, Richard, 15
Poster, Marc, 141
Pounder, Chris, 121–122
Prosser, William, 210, 218

Raab, Charles, 120, 122–123
Radin, George, 31
Rankin, Murray, 142
Rataczack, Stanley, 307–309
Reagan, Ronald, 170
Regan, Priscilla, 2, 120, 123, 215
Reich, Robert, 303
Reidenberg, Joel, 122
Reiman, Jeffrey, 66–67
Reingruber, Michael, 40–41
Reinig, Bruce, 68
Rifkin, Jeremy, 162
Rivest, Ronald, 254–255
Rogers, Martha, 304
Rogers, Paul, 30
Root, Robert, 69
Rosenthal, Arnon, 17
Rotenberg, Marc, 15, 123
Roy, Jean-Hugues, 307
Ruggles, Myles, 305
Rule, James, 94, 303

Sabel, Charles, 303
Samarajiva, Rohan, 4, 8–11, 19, 22, 64, 67, 304–305, 309
Samuelson, Pamela, 304
Sandel, Michael, 240
Sankey, William, 305

Sasse, Ulrich, 237
Scher, Julia, 141
Schmandt, Christopher, 66
Schneier, Bruce, 141, 248, 274–275
Schoeman, Ferdinand, 1, 25
Schuster, Neil, 165
Schwartz, Alan, 2
Schwartz, Evan, 304
Schwartz, Paul, 109, 121–122, 211, 215, 218
Sciore, Edward, 17
Senge, Peter, 142
Shamir, Adi, 254–255
Shapiro, Susan, 244
Sharp, Leslie, 152
Shepard, David, 304
Siegel, Michael, 17
Simitis, Spiros, 121, 141, 177–179, 214, 218, 237–240
Simon, Herbert, 286
Simon, Paul, 162
Simsion, Graeme, 41, 49–51
Smith, H. Jeff, 2, 123, 237, 305
Smith, Randy, 70
Smith, Robert, 1, 122
Smythe, Dallas, 303, 306–307
Star, Susan, 18, 48
Stearnes, Linda, 303
Steinberg, Steve, 304
Steinmuller, Wilhelm, 141, 238
Sterling, Bruce, 47
Stewart, Potter, 204
Stone, Eugene, 67
Stults, Robert, 69
Suchman, Lucy, 49

Tang, John, 69, 76
Tapscott, Don, 25, 123
Taylor, Frederick, 33
Tenner, Edward, 164
Thatcher, Margaret, 170
Thivierge, Robert, 307–308
Thompson, Paul, 33
Trudel, Pierre, 307–309
Turkington, Richard, 7
Turow, Joseph, 306–307

Uglow, David, 303

Van Koughnett, Greg, 308
van Stokkom, Bas, 141
Varian, Hal, 305
Vogel, Harold, 306–307

Walkerdine, Valerie, 44
Walter, Helmut, 238
Want, Roy, 65
Warren, Samuel, 203, 217, 305
Weiner, Tim, 255
Weisband, Suzanne, 68, 93
Weiser, Mark, 65
Weiss, Michael, 303
Westin, Alan, 1–2, 123, 162, 164,
    194, 216, 240, 308
Wildman, Steven, 306
Williamson, Oliver, 13
Wippermann, Gerd, 241
Wright, Charles, 307

Yates, JoAnne, 35

Zimmerman, Phil, 246, 265–266
Zureik, Elia, 1

Subject Index

Activism, 1, 4, 21, 31, 45, 55, 105–106, 114, 143–144, 155, 162, 170, 185–186, 246, 262–263, 266, 270, 273, 301–302, 309

Advertising, 15, 134, 147, 277, 280, 288–289, 292, 297, 307

Airlines, 110

Alberta, 168, 173

Anonymity, 10, 52, 54, 127–132, 135–136, 138, 160, 245, 253–254, 258–259, 262, 266–267, 271–274, 279, 297

Apple, 71–72, 81–83, 87–88, 92

Architecture, of computer systems, 3–5, 11–12, 21–24, 30–32, 53

Artificial intelligence, 40–41

Assembly, freedom of, 146, 149, 193

Association, freedom of, 11, 26, 31

AT&T, 264, 288

Attention, 3, 22, 130, 136, 285–289, 292–294

Audiences, 22, 278–280, 286–296, 300–301, 306

Audio, 65–70, 83–86

Auditing, 25, 53, 115–116, 175, 178, 184–186, 244

Australia, 100–101, 113, 146–149, 153–159

Australian Privacy Charter, 155

Australian Privacy Foundation, 147, 155

Austria, 100, 112, 221, 223, 226, 228, 231, 233, 237, 239–240

Authentication, 48, 53–55, 72, 82, 86

Autonomy, 31, 187

Badges, active, 65, 74

Banks, 9, 56, 126–127, 151–152, 205, 209–212, 228, 245, 247, 254, 257–258, 266–271, 275, 292, 296, 299

Bargaining, 12, 106, 212, 232–233

Bavaria, 222

Belgium, 233

Borders, flows of data across, 5, 80, 107–110, 220, 227, 236, 239

Boundaries
jurisdictional, 3–4, 21, 259
legal, 202
organizational, 17, 53
personal, 8–11
public-private, 45, 63, 73, 89
technical, 21

British Columbia, 157, 168, 170–190

British Columbia Civil Liberties Association, 178, 180

California, 9, 29–30, 53, 55, 72

Caller ID, 8–9, 55, 144, 146, 161

Cameras, 11, 43, 65, 69–73, 82–92, 149–152, 164

Canada, 100–101, 111–117, 120–121, 155, 167–168, 173–174, 178, 180, 183–185, 190, 291–294, 299. *See also names of provinces*

Canada Post, 291, 295–296

Canadian Privacy Council, 154
Canadian Standards Association, 25,
   111, 115–116, 154, 189, 215
Capture, 18, 30, 48, 66, 69–70, 77,
   79, 81, 83, 87–88, 91–92
Categorization, 16, 18, 35, 42, 48–50
Censorship, 146
Censuses, 16, 154, 164, 222
Client-server systems, 21
Clipper, 23, 114, 117, 155, 164,
   245–246, 258–266
Codes of practice, 2, 7, 10, 12,
   24–25, 80–81, 110–111, 114–115,
   118, 123, 144, 152, 189–190,
   215, 241, 244, 294–300
   sectoral, 24, 100
Coercion, 11–12, 284–285, 300, 302
Commodities, 16, 103, 144, 153,
   160–161, 174
Communication, computer-mediated,
   64, 66, 68, 75–76, 78, 90–91
Communications technology, 1–4, 19,
   31, 57, 63–65, 68, 102, 104, 131,
   137, 140, 230, 236, 257, 263,
   267, 273, 277–279, 285, 289, 298
   wireless, 3, 30, 65, 277, 280
Compuserve, 289
Computer Professionals for Social
   Responsibility, 264
Computer science, 40, 42, 57
Computer-supported cooperative
   work, 66–68, 74–78, 90
Computing, 104
   distributed, 17, 42, 65, 69, 262
   multimedia, 63–65, 68, 71
   ubiquitous, 65–69, 74, 77–79, 84,
   90–93
Confidentiality, 44, 183, 212,
   217–218
Connecticut, 174
Consent, 80, 104, 126, 128, 156,
   198, 217, 227, 230, 232, 234,
   236, 278, 283–284, 298, 300
Consumers, 15, 24, 105–106, 111,
   116, 130, 143–147, 155, 158,
   160, 210–212, 215, 227, 270,

277–279, 282–285, 290–291,
   294–297, 301
Contests, 3
Contracts, 11, 14, 17, 24, 109,
   212–214, 232–233, 236
Copyright, 281
Correction, 104, 186, 197, 202, 211,
   217, 226, 233
Council of Europe, 99, 105–106, 109,
   156, 220
Credentials, 53
Credit cards, 56, 126–127, 131, 160,
   268–271, 289, 292, 296
Credit reporting, 115, 209–210, 233,
   241
Cryptography, 4, 9–11, 17, 23, 32,
   53–56, 77, 117–118, 125–126,
   133, 146, 164, 243–275
Customization, mass, 277–279, 300,
   303
Customs, 12–14, 55, 208

Data banks, 102–103, 170, 204–205,
   221–227
Data collection, 3, 16, 80–81, 84,
   103, 115, 117, 129, 144, 156,
   169, 173, 194–199, 210–211, 222,
   254, 277–281, 293, 296–298
Data interchange, 130
Data matching, 103, 105, 144, 146,
   170, 175–176, 187, 198–200
Data mining, 3
Data modeling, 16–17, 40–42, 50, 52
Data protection, 1–7, 10, 12, 14, 16,
   21, 31, 52, 56, 99–104, 107–120,
   129, 136–137, 143–144, 156–158,
   161, 167–179, 182, 185–190,
   194–195, 219–237
Databases, 2–4, 8, 16–18, 23, 39–42,
   45, 48–57, 93, 103, 145–146, 158,
   210, 259, 261, 280–281
   merger of, 3
Defaults, 9, 24, 71, 78, 89, 117
Democracy, 10, 26, 43–45, 112,
   136–138, 153, 171
Denmark, 100, 220, 226–228, 233,

237, 239, 241

Data Encryption Standard, 245, 254–258, 266, 275

Digicash, 247, 268–272, 275

Digital cash, 23–24, 247, 254, 268

Digital Equipment Corporation, 257

Disclosure, 8, 12, 56, 80, 90, 105, 161, 169, 172, 177, 194–195, 198–199, 202–205, 210–212, 217, 244, 282–283

Disembodiment, 73–76, 89–90

Dissociation, 73–76, 82, 86

DNA testing, 146, 159

Drug testing, 156, 184

Education, 9, 25, 55, 117–118, 120, 147, 172, 175, 292, 301

Electronic Frontier Foundation, 263–266

Electronic mail, 4, 128, 164, 180, 207, 212, 258–259, 289, 292, 297

Electronic Privacy Information Center, 155, 164, 174

Enforcement, 101, 109–110, 114–115, 118, 152, 168, 197, 200–202, 209, 213, 215, 227–228, 233–234, 293–295, 300

England, 52, 159, 163

Entity-relationship model, 17, 39, 50

Environmental protection, 119

Europe, 2, 100–101, 105–106, 108, 110–114, 118, 131, 136, 143, 145, 148, 154, 156, 158, 164, 172, 194–195, 219–222, 235–236, 247, 293

Council of, 99, 105–106, 109, 156, 220

European Union, 5, 52, 104–119, 162, 189, 214, 218, 221, 233–236

Export, 245, 256, 258, 264, 267, 273

Expression, freedom of, 149, 193

Fair information practices, 2, 23–24, 52, 99, 101, 104, 106, 113, 116, 119, 156, 170, 175, 183, 186–189, 194–197, 200–202, 205–215

Federal Bureau of Investigation, 114, 246, 263–264, 267

Feedback, 8, 57, 66, 68, 70–73, 76–87, 91–94

Fingerscanning, 54–55

Finland, 231–234, 237

France, 67, 100, 162, 167, 220, 226, 228, 233, 237, 294

France Telecom, 294

Fraud, 22, 147, 199

Freedom of Information and Privacy Association, 178, 180

Genetic analysis, 2, 162

Geography, 3, 16, 48, 132–133, 144, 174

Germany, 100–101, 112, 136, 161–164, 167, 177, 185, 221–224, 227–239. *See also names of states*

Globalization, 1, 5, 19, 23, 99, 121, 273

Greece, 108, 113

Groupe Videotron, 291

Harmonization, of data-protection policy, 22, 100, 105–109, 118–119, 218, 297

Hawaii, 174

Hearst Interactive Canada, 291, 295

Hesse (German state), 136, 177, 214, 221–224, 240

Hollerith cards, 35–36

Hong Kong, 110

Hungary, 101, 109, 112, 162

Hydro-Quebec, 291, 295

IBM, 255

Identification, 16, 18, 29, 48, 51, 55, 67, 86, 126–127, 131–132, 135, 145–147, 153, 155, 159–160, 163, 173, 251–254, 266, 270–271, 274

biometric, 54–55, 160

Identity, 1, 3, 7–10, 48, 52–57, 76–77, 117, 125, 131, 137–138, 146, 154, 170, 177, 210

theft of, 22

Industry associations, 2, 116, 118, 150
Information, freedom of, 131, 153, 162, 164, 168–169, 174, 181, 201
Information technology, 2, 19, 63–64, 99, 102–104, 117, 140, 145, 154, 170, 175, 187, 195, 199, 211, 219, 230, 236, 243, 278–279, 286
Insurance, 110, 147–148, 189, 209–210, 292
Intellectual property, 127
Internet, 4–5, 17, 21, 31, 138, 146, 155, 164, 190, 235, 246, 264, 266–270, 273, 277, 287–288
Interoperability, 17, 258
Ireland, 240
Italy, 108

Japan, 100, 113
Jurisdiction, 99, 110, 169, 214, 245, 251, 253, 262–263, 266–267, 273–274, 299

Law, 1, 3, 5–7, 12, 21, 24, 29–30, 52, 64, 67–68, 80–81, 91, 100, 102, 104, 108–114, 118–119, 128, 136, 140, 143–144, 149, 168–174, 179, 193–202, 207–210, 213–214, 221, 244, 258, 270
administrative, 101
civil, 299
common, 25, 195, 209, 211, 218, 299
constitutional, 7, 100, 193, 202–208, 213, 226, 229, 231, 234, 257
enforcement of, 4, 117, 149, 151, 156, 162, 185, 207–208, 217, 253, 259–260, 263, 267, 270–272
Legislation, 10, 25, 48, 67, 99–101, 104–116, 120, 136, 145, 147, 157, 162, 171–172, 176, 185, 189, 196, 198, 207, 210, 214, 222–223, 230, 235–236, 297, 301
Legislative process, 2, 48, 106, 114, 172, 221, 231

Legislatures, 24, 104, 168–171, 174, 178–182, 194, 199, 222–224, 227, 230, 234, 236, 239, 245, 247, 263, 273
Litigation, 174, 197, 200–201, 206, 209–212
Loto-Quebec, 291–292, 295
Lotus, 164
Luxembourg, 240

Magazines, 15, 289
Manpower (employment agency), 148
Manufacturing, 1, 19, 33–36, 263, 279, 290
Marketing, 2, 9, 19, 134, 232, 257, 271, 278–281, 286, 288, 292–295, 301, 307
direct, 110–111, 115, 134, 187, 210, 227, 234, 241, 288
Massachusetts Institute of Technology, 246, 265, 268
Mastercard, 268, 271
Media institutions, 137, 155, 157, 163, 169, 172, 175–178, 180, 182, 277–278, 280–281, 286, 288, 291, 294
Media spaces, 8–9, 69–76, 82, 89–92
Media technologies, 1, 4, 26, 32, 39, 71, 284–285, 287, 289, 300, 302
Medicine, 2, 45, 115, 133, 146, 153, 158, 162, 183, 186, 204–205, 214
Microphones, 65, 69, 73, 91
Minitel, 294
Minnesota, 174
Mirror Worlds, 42–49
Mobilization, 6, 19, 22, 134–136, 138, 246, 267, 273
Money laundering, 267, 271
Motion study, 35
Movement, freedom of, 146, 149
Mytec Technologies, 54

National Bank of Canada, 291–292, 298–299
National Bureau of Standards, 255, 258

National Science Foundation, 256
National Security Agency, 255–258
Negotiation, 8–16, 20, 24, 109, 115, 232, 252–253, 263, 266, 270, 272, 283
Netherlands, 22, 52, 100, 114–115, 140, 154, 157, 231, 240, 271
Network externalities, 17, 19, 290, 292, 300
Networking, computer, 4, 13, 16, 20–21, 24, 31, 65, 68–69, 103–104, 158, 188, 214, 230, 243, 250, 263, 268, 272–273, 301
New Brunswick, 294
Newspapers, 289
New York, 174
New Zealand, 110–115
New Zealand Privacy Foundation, 154
Norway, 100, 226–228, 231, 233, 237, 240
Notification, 12, 24, 107, 218

Online services, 4, 8–9, 190, 212, 294
Ontario, 22, 52, 55, 140, 168, 173, 180
Organization for Economic Cooperation and Development, 80–81, 93, 99, 101, 105–106, 109, 113–116, 121, 156, 220

Paperwork, 35–36, 39, 51, 65, 197
Participation, 135–138, 229–236
  individual, 81
Passports, 160
Patents, 245–248, 255, 259, 262, 265, 271
Persian Gulf War, 47
Persona, digital, 3, 7, 138
Photography, 54, 146, 170, 203
Police, 7, 55, 131, 146, 149–154, 157–160, 164, 178, 183, 185–186
Polls, 6, 117, 143–148, 163
Portholes (system), 72, 82, 85, 88, 92
Portugal, 112, 162, 226, 240
Pretty Good Privacy, 245–246, 258–265, 267, 275

Privacy commissioners, 5, 21–24, 52, 101, 105, 109, 157, 168–169, 173–178, 182–189, 227–228
Privacy-enhancing technologies, 4, 6–7, 9–10, 20, 23–25, 52, 55–56, 117, 125–138, 140
Privacy International, 105, 155, 157, 164
Privacy rights, 2, 6–7, 25, 101, 144, 147, 152, 155, 160, 164, 197, 202–205, 217–218, 225–226, 235
Process charts, 33–35, 38–39
Profiling, 103, 105, 153, 170, 188, 210, 281, 295
Pseudoidentity, 53–56, 125
Public health, 133, 186
Publicity, 172, 210
Public relations, 15, 182, 246–247
Public space, 8, 10–11, 26, 43–44, 63, 69–74, 82, 85, 87, 89, 91, 149–152, 164, 283
Public sphere, 4, 11, 19, 45

Quebec, 8, 22, 110, 116, 168, 173, 189, 290, 294, 297, 299, 301

Radio, 29, 172, 286, 289
RAVE (Ravenscroft Audio-Video Environment), 69–75, 81–93
Records, 13, 39–40, 42, 47, 50–57, 127, 186–187, 193–199, 201, 212–215
  cable television, 202
  census, 222
  criminal, 93, 170, 174, 176, 181, 210
  educational, 202
  financial, 93
  manual, 107
  medical, 45, 93, 183, 188, 212–213, 304
  order, 35
  public, 211
  telephone, 4
  video rental, 202, 304
  voter registration, 205, 210

Registration, 101, 104, 107, 116,
223–228, 231, 239
Remailers, 245–246, 258–259,
261–263, 266–268, 271, 275
Reputation, 13, 15
Rheinland-Pfalz (German state), 221,
224, 240
Rivest-Shamir-Adleman algorithm,
245, 254–258, 265

Satellites, 29, 156, 280, 288
Scandinavia, 162
Scotland, 152
Seclusion, 7, 210
Secondary use, 16, 22, 53, 56–57
Secrets, 7, 11, 131, 194, 222, 243,
253, 258, 262, 272–273
Secure Electronic Transaction, 247,
268–271, 275
Security
computer, 17, 44, 53, 57, 65, 68,
80, 109, 118, 125, 130–131, 136,
139, 178, 188, 200, 205–206,
218, 223–224, 230, 236, 243,
249–250, 253, 257, 297, 299
national, 162, 164, 200
personal, 11, 149, 281
public, 137, 152, 162, 164
Self-determination, 31, 229–235
Sensors, 3, 49, 75
ShrEdit (text editor), 75
Sierra Research, 30
Signatures, digital, 117–118, 125,
131, 250, 255–256, 261
Smart cards, 54, 117, 153, 160, 292,
297
Social control, 4, 31, 44, 58, 88, 151,
244
Software, 4, 42, 45, 49, 65, 69, 89,
127, 257, 264, 267
Solitude, 149, 210, 283
Spain, 112, 226, 240
Standards, 17–18, 24–25, 100, 111,
115, 117, 155
administrative, 5, 16, 116, 123,

152, 154, 164, 177, 194, 253
legal, 14, 115, 198, 200, 205, 211,
214, 230
technical, 16, 21, 23, 48, 65, 118,
255–258, 264, 266, 274
Stichting Waakzaamheid
Persoonsregistratiie (Netherlands),
154–155
Subpoenas, 217, 253, 273
Surveillance, 2, 11, 26, 30–31, 44,
47, 49, 64, 91, 103, 139–140,
143–152, 156, 159, 161, 164,
168–172, 175, 181–184, 207, 223,
243, 255, 263–265, 273, 278–285,
288–291, 294, 296, 300–302
Sweden, 52, 100–101, 167–168,
221–224, 237, 240–241
Switzerland, 233–234, 237
Systems analysis, 31, 35, 39–40, 57

Tabulating machines, 35, 38
Taxes, 146, 156, 160, 162, 212, 218,
222, 270
Telephones, 4, 7–9, 11, 21, 56, 76,
151, 180, 202, 207, 259, 264,
280–281, 288–289, 294, 297, 305
cellular, 93
Teletel, 294
Television, 43, 280, 286–289, 293
cable, 202, 277, 280–281, 288,
292–293, 305
closed-circuit, 146, 150–152, 157
Thailand, 220
Therblig (unit), 35–36
Toll collection, 14, 30, 56, 160
Torts, 25, 209–214
Tracking, 3, 19, 26, 30, 65, 68, 144,
153, 222, 271, 289
Transaction costs, 13–17, 20, 110,
236
Transparency, 6, 12, 15, 22, 56, 104,
131, 134, 172, 196, 271
Transportation, 3, 13–14, 19, 29–30,
40–41, 53, 151, 153, 189
Trust, 10–11, 25, 55, 68, 71, 74, 79,

83, 86–88, 104, 126, 137–140,
  145, 152, 243–245, 248, 251–255,
  258, 261–263, 266–267, 270–274,
  282–285, 290–293, 296, 300–302
Turkey, 113

UBI, 289–301, 307–308
United Kingdom, 11, 67, 72,
  100–101, 105, 107, 112, 115,
  145, 150, 152, 156–158, 220,
  237–240. *See also* England;
  Scotland; Wales
United States, 5, 23, 47, 55, 67, 101,
  105, 113–118, 120–121, 131, 138,
  145–148, 158, 160, 167–168, 173,
  187, 193, 195, 197, 209,
  212–215, 220, 226, 236, 245,
  247, 253–259, 262–263, 273–275,
  281, 288–291, 293, 299–300. *See
  also names of states*
Usenet, 265

Video, 8, 11, 43, 48, 65–71, 73, 77,
  82–88, 93, 127, 151, 264, 292
Videocassette recorders, 280, 290
Videotex, 294
Videoway Multimedia, 291–292
Virginia, 205, 210
Virtual Café, 72, 74, 81–88, 91–92
Virtual reality, 47, 90
Visa, 268, 271
Voicemail, 289
Voluntariness, 12, 110, 115, 118,
  143–144, 159–160, 264, 280, 283
Voting rights, 206

Wales, 151, 163
Warrants, 7, 207–208
Welfare state, 2, 6, 55, 170, 222, 228
Wiretapping, 4, 114, 203–208,
  212–213
Workplaces, 2, 8, 67, 134

Xerox, 69, 72